D1824654

Environmental Hazards

Series editor
Thomas A. Birkland

More information about this series at http://www.springer.com/series/8583

Fernando I. Rivera • Naim Kapucu

Disaster Vulnerability, Hazards and Resilience

Perspectives from Florida

 Springer

Fernando I. Rivera
Department of Sociology
University of Central Florida
Orlando, FL, USA

Naim Kapucu
School of Public Administration
University of Central Florida
Orlando, FL, USA

Environmental Hazards
ISBN 978-3-319-16452-6   ISBN 978-3-319-16453-3   (eBook)
DOI 10.1007/978-3-319-16453-3

Library of Congress Control Number: 2015936283

Springer Cham Heidelberg New York Dordrecht London
© Springer International Publishing Switzerland 2015
This work is subject to copyright. All rights are reserved by the Publisher, whether the whole or part of the material is concerned, specifically the rights of translation, reprinting, reuse of illustrations, recitation, broadcasting, reproduction on microfilms or in any other physical way, and transmission or information storage and retrieval, electronic adaptation, computer software, or by similar or dissimilar methodology now known or hereafter developed.
The use of general descriptive names, registered names, trademarks, service marks, etc. in this publication does not imply, even in the absence of a specific statement, that such names are exempt from the relevant protective laws and regulations and therefore free for general use.
The publisher, the authors and the editors are safe to assume that the advice and information in this book are believed to be true and accurate at the date of publication. Neither the publisher nor the authors or the editors give a warranty, express or implied, with respect to the material contained herein or for any errors or omissions that may have been made.

Printed on acid-free paper

Springer International Publishing AG Switzerland is part of Springer Science+Business Media (www.springer.com)

Dedicated to

Fernando: wife Lymari, and children Ernesto, Mariana, and Daniela as they are my guiding light and inspiration. Special dedication also goes to my mother, who unexpectedly passed away while I was writing this book.

Naim: wife Ayşegül and children Emre, Selim, and Yusuf

We also want to dedicate this book to all emergency personnel who are at the forefront of building resilient communities. Finally, we dedicate this book to all underserved and vulnerable groups; we hope this book provides them with a voice and a place at the table as we strive to continue to build disaster-resilient communities.

Foreword

Increasing exposure to hazards and increasing social and economic vulnerability are raising the specter of catastrophic disaster in the United States. Florida is the perfect example as hurricanes and other forms of severe weather, drought and wildfire, agricultural and human diseases, the threat of foreign and domestic terrorism, and other natural and unnatural hazards are increasing the risk to life, property, and the environment. When we think of natural hazards in the United States, our first thoughts typically are of earthquake risks in California and hurricane risks in Florida. While other states are faced with those and other hazards, recent experience has demonstrated the certainty of those risks and high costs. The risk of catastrophic disaster is probable, and the risk of significant losses is certain.

Tropical cyclones have affected about half of the American states, but Florida has been the most frequent target of the storms, particularly the catastrophic storms. Indeed, Florida has a very long history of severe storms, and population growth and concentration is raising the ante in terms of potential losses. Growing social vulnerability is exacerbating the risks, as well. The lessons learned from hurricanes in the last century may not be remembered as clearly as they should be, but lessons learned in the last several decades may still provide impetus for preparedness and mitigation programs. For example, the Great Labor Day Hurricane of 1935 which caught railroad workers, many WWI veterans, exposed in the Florida Keys, demonstrated the importance of early warning. The science of cyclones was not well understood at the time, and little was done to prepare for the possibility of storms. Hurricane Betsy in 1965, Hurricane David in 1979, and Hurricane Andrew in 1992, as well as dozens of lesser storms, also focused attention on the needs for better warning systems, building codes and standards, strategies for evacuation, and plans for long-term recovery. The value of vertical evacuation, the need to focus on populations with functional needs, and the necessity of improved logistics were not on the emergency managers' radar at that time. The use of social media for warning and response operations was a topic more likely to appear in science fiction books. Clearly, much of what we know about hurricanes, including the patterns of destruction wrought by wind and water, we gleaned from experiences in Florida with those early storms.

Riverine flooding in the Midwest and West, coastal flooding on the East coast, tornadoes in the Midwest and elsewhere in the South, wildfire in the Mountain states and West, terrorist attacks in New York City and Tulsa, and other disasters similarly provided lessons for Florida. Emergency managers, public health officials, private sector risk managers, and other professionals are increasingly applying the lessons learned all over the United States, as well as in the rest of the world, to prevent losses of life, property, and environmental quality. In perhaps lesser measure, communities are being forced to adopt appropriate and effective land use, zoning, and building regulation to reduce losses – or face legal liability for failures to act responsibly to protect their residents. Also slowly, states and local governments are having to consider their legal exposure when hazardous industries are not effectively regulated to prevent accidental disasters. Recent failures to deal with known hazards range from massive explosions to life-threatening air, ground, and water contamination. Pressure is also increasing for community and state action to address newly identified hazards, including climate-related diseases affecting agriculture and human beings. Florida officials have already noted the increasing incidence of dengue fever and other mosquito-borne diseases.

Communities have become much more vulnerable as population density has risen along the state's Atlantic and Gulf coasts and in the Keys. The vulnerability of the population has increased as well with influxes of retirees, agricultural workers, international and domestic tourists, and part-time residents. The impacts on infrastructure, the health-care system, the educational system, and social services are one thing, but the additional stress on critical services during disasters is significant. In a low-tax state, essential services are fiscally strained even in the best of circumstances. What is changing is that the vulnerability of people, property, and the environment is increasing. The impacts of climate change and sea level rise are already being felt in terms of higher storm surges, even for relatively minor storms, and higher wind speeds. While scientists are being careful about attributing those increases to climate change, they have ample evidence that the increases are taking place.

As the annual estimates of tropical storms and hurricanes and the expected landfalls are made, attention often shifts to Florida. State and local emergency management agencies begin assessing capabilities and reviewing plans. They watch the tropical disturbances developing off the coast of Africa and hope that one or more of the disturbances will not become the next catastrophic disaster to strike along the coast. Such is life in Florida and elsewhere on the Gulf and up the East coast. Hurricane season is stressful, and it is a stress that emergency managers in Florida are accustomed to.

Fortunately, the State of Florida has an exceptional emergency management system. A major factor in that success has been the level of state support for local agencies. State encouragement of locals to meet national standards and to seek Emergency Management Accreditation Program (EMAP) accreditation has also been a factor. The EMAP Standard focuses attention on the whole emergency management program, including all stakeholders involved in disaster planning and operations, and the development of a comprehensive program. Collaboration is critical. But, still, emergency management agencies in Florida and elsewhere in the United States have uneven capabilities. Some are very professional and as prepared to deal with hurricanes and other hazards as they can be, while others are barely able to contend with the everyday minor disasters caused by fire and flood. Volunteers, nonprofit organizations, and private firms provide surge capacity and help fill in the gaps.

It should be noted that changes are taking place in emergency management at the policy level. The cavalry approach, relying on federal and state officials to provide essential resources, is quickly shifting to a more self-sufficient approach, asking individuals, families, and communities to take more responsibility for their own health and safety. There are a lot of reasons for the shift in policy and program, including the expense of large-scale disaster operations and long-term recovery, but the most important reason is that catastrophic disasters tax the capabilities of the public, private, and nonprofit sectors. As the Katrina and Sandy disasters demonstrated, help may be slow in coming. Indeed, there may be catastrophic disasters in which help may be weeks or months away and communities may have to take care of themselves for a significant length of time. The response to Hurricane Katrina came together very slowly and overwhelmed state and local capabilities. Pandemics can overwhelm even national capabilities, as the current Ebola crisis in West Africa demonstrates. But major earthquakes, volcanic eruptions, tsunamis, and hurricanes can sorely test emergency management capabilities at all levels. This is why the emphasis in recent years has been on community resilience, making the community less vulnerable and less reliant upon outside resources and enabling it to recover quicker. The goal is to replace infrastructure and to make it less vulnerable to the next storm in order to help communities find a new normal. While there is still no consensus on the definition of resilience, there is growing understanding of its key elements.

I, too, keep an eye on the tropical disturbances as they form off the coast of Africa. I watch as the storms move close to my mother's home in Largo and pay attention to the vulnerability of Largo and other communities around Tampa Bay. I pay attention to the quality of the professional emergency managers hired by the State of Florida and Pinellas County. I applaud the adoption of measures to reduce risks in and around Largo and to make the community more resilient. This is more than an intellectual exercise. It is a matter of preparedness for risks that might affect my family and require quick action on my part.

Georgia State University   William L. Waugh Jr.,
Atlanta, GA, USA

Preface

Several disasters, such as four hurricanes within a 6-week period in Florida of 2004, Hurricane Katrina in 2005, and Super Storm Sandy in 2012, provided unfortunate reminders of the vulnerability of communities to natural disasters. These disasters, like many others, illustrated how events harm communities and individuals and disrupt social-technical systems and community functions. Florida is one of the United States' most at-risk states for disasters. Weather patterns make many regions of the state vulnerable to hurricanes. Droughts often produce wildfires, and unstable weather conditions can create tornadoes. Partly due to these constant disaster threats, the Florida emergency management system is ranked among the nation's best (Jordan 2006), and it has been identified as a model for the entire United States (Waugh 2006).

The academic debates on the concept of resilience are plentiful (National Research Council 2010), yet few studies have explored the perception for the personnel responsible for disaster emergency management and response. This book fills this gap by providing an analysis of the narratives from individuals representing different emergency management sectors, including county emergency managers and nonprofit and community groups.

In addition, the book explores these issues in rural community settings, an often neglected sector of study in the disaster literature. The capacity of rural communities to plan for and respond to disasters, in particular, has been the focus of recent studies (Cutter et al. 2010; Kapucu et al. 2013; Skerratt 2013; Whitman et al. 2013). Compared to urban areas, rural communities may have a less diversified economic base and fewer financial resources to support disaster mitigation practices or rebuilding efforts. Thus, an emphasis of rural emergency management adds to this growing of the disaster research literature.

This book offers a unique examination into the perceptions of disaster vulnerability and resilience. It will contribute to the literature on emergency management, policy, and planning as it pushes the resilience and vulnerability debate from the theoretical to the actual practical realm of the perceptions of what is currently happening in disaster management groups and organizations. It will also contribute to the sociological literature as it analyzes the disaster emergency personnel perspectives on vulnerable groups, such as the elderly, children, and farmworkers, among others. Ultimately, this book sheds new light on how disaster emergency personnel perceive disaster resilience and vulnerability and provides evidence on how perceptions sometimes can shape reality – a reality separate from established policies and governance mechanisms.

Key Elements

The role of culture, social capital, socioeconomic vulnerabilities, and interpersonal social networks provides complementary evidence to the analyses conducted at the larger community and regional scale of disaster planning and management. The relationships among land uses, housing decisions, and mitigation strategies addressing the vulnerability to disasters. The literature on disaster mitigation, preparedness, recovery, and economic development casts light on the challenges for communities for creating an integrated disaster preparedness and making development efforts. This refocuses attention away from a "silo" approach to a "collaboration" approach in disaster-resilient and sustainable communities. These perspectives will lead to the development of strategies for improved management in the mitigation, preparation, response, and recovery to/from natural hazards.

Research Questions

The following questions are not intended to be exhaustive of the propositions examined in this book, but indicate some basic questions we seek to investigate with contributing scholars in the field:

1. How can the concept of resilience be operationalized/used in a way that is useful as a framework to investigate the conditions that lead to stronger, safer, and more sustainable communities?
2. What factors account for the variation across jurisdictions and geographic units in the ability to respond to and recover from a disaster?
3. How does the disaster recovery process impact the social, political, and economic institutions of the disaster-stricken communities?
4. How the disaster-impacted communities, especially rural ones, collaborate with multiple stakeholders (local, regional, state, national) during the transition from recovery to resilience?
5. Can the collaborative nature of disaster recovery help build resilient communities?

The book offers a systematic, empirical examination of the concepts of hazards, vulnerabilities, and disaster resilience, focusing on communities in Florida.

The findings bring new evidence and insights into the study of disaster resilience, integrate knowledge from sociology and public policy and governance to the study of natural disasters, and provide useful and accessible insights into academic circles along with the government, nongovernmental organizations, and the private sector. The results are generalizable and widely applicable to a variety of circumstances throughout the world.

The primary audiences of this book are scholars in emergency and crisis management, planning and policy, disaster response and recovery, disaster sociology, and environmental management and policy. This book can also be used as a textbook in graduate and advanced undergraduate programs/courses on disaster management, disaster studies, emergency and crisis management, environmental policy and management, and public policy and administration. The book can also be useful for scholars and students in different regions of the world.

Acknowledgments We acknowledge guidance and assistance of Thomas A. Birkland, Environmental Hazards series, and Fritz Schmuhl, Springer editor. We also acknowledge assistance from Judith Terpos during the production phase of the book and comments from anonymous reviewers of the book project. In addition, our student assistants Brittany Haupt provided research and literature support and Sergio Ramirez assisted in content analysis. The data utilized in this book comes from a study supported by the U.S. Department of Agriculture under Award No. 2010-67023-21698. Any opinions, findings, and conclusions or recommendations contained in this document are those of the authors and do not necessarily reflect the views of the agencies that provided support for the project. Finally, we would like to thank all of the government officials and nonprofit, community, and faith-based organizations who participated in the study and were gracious enough to share their opinions and insights into developing more disaster-resilient communities.

Orlando, FL, USA   Fernando I. Rivera
                                                                                           Naim Kapucu

References

Cutter, S. L., Burton, C. G., & Emrich, C. T. (2010). Disaster resilience indicators for benchmarking baseline conditions. *Journal of Homeland Security and Emergency Management, 7*(1), 1–24.

Jordan, L. J. (2006, February 15). *Report slams Katrina disaster preparation.* Associated Press Wire Service.

Kapucu, N., Hawkins, C., & Rivera, F. (2013). *Disaster resilience: Interdisciplinary perspectives.* New York: Routledge.

National Research Council (NRC). (2010). *Building community disaster resilience through private-public collaboration.* Washington, DC: The National Academies Press.

Skerratt, S. (2013). Enhancing the analysis of rural community resilience: Evidence from community land ownership. *Journal of Rural Studies, 31,* 36–46.

Waugh, W. L. Jr. (Ed.). (2006). *Shelter from the storm: Repairing the national emergency management system after Hurricane Katrina.* Special issue of *The Annals of the American Academy of Political and Social Science,* 256–272.

Whitman, Z. R., Wilson, T. M., Seville, E., Vargo, J., Stevenson, J. R., Kachali, H., & Cole, J. (2013). Rural organizational impacts, mitigation strategies, and resilience to the 2010 Darfield earthquake, New Zealand. *Natural Hazards, 69*(3), 1849–1875.

Contents

About the Authors

Fernando I. Rivera, Ph.D. is an Associate Professor of Sociology at the University of Central Florida. His research interests and activities fall under staple areas of sociology, primarily the sociology of health/medical sociology, disasters, and race and ethnicity. In the area of health, his published work has investigated how different mechanisms are related to certain health and mental health outcomes with a particular emphasis on Latino populations. His disaster research has explored the investigation of factors associated with disaster resilience, particularly in rural communities. In the area of race and ethnicity, he has investigated the effects of segregation and perceived discrimination on the health of Latino populations. Dr. Rivera has received funding from the U.S. Department of Agriculture for his disaster research. In addition, he participated in the Research Education in Disaster Mental Health training fellowship from the National Institute of Mental Health and Dartmouth University. He is also the editor of *Disaster Resilience: Interdisciplinary Perspectives* (2013, with Kapucu and Hawkins). He earned his M.A. and Ph.D. in Sociology from the University of Nebraska–Lincoln and his B.A. degree in Sociology from the University of Puerto Rico–Mayagüez. He also completed an NIMH-sponsored postdoctoral fellowship at the Institute for Health, Health Care Policy and Aging Research at Rutgers, the State University of New Jersey.

 **Naim Kapucu, Ph.D.** is Director and Professor of Public Policy and Administration at the School of Public Administration at the University of Central Florida (UCF). He is the founding Director of the Center for Public and Nonprofit Management (CPNM) at UCF (2008–2011). He also directed the Master of Public Administration (MPA) program. He has developed Emergency Management and Homeland Security graduate certificate and undergraduate minor programs at UCF. He has published widely in the areas of public policy and administration, crisis leadership, nonprofit management, and disaster management. His work has been published in *Public Administration Review*, *Administration & Society*, *Journal of Public Administration Research and Theory*, *The American Review of Public Administration*, and *Disasters*, among many others. His main research interests are emergency and crisis management, network leadership and governance, decision-making in complex environments, and collaborative governance. He is the author of *Network Governance in Response to Acts of Terrorism: A Comparative Analysis* (2012) and *Managing Emergencies and Crises* (2013, with Alp Özerdem). He is the editor of *Disaster Resilience: Interdisciplinary Perspectives* (2013, with Hawkins and Rivera) and *Disaster and Development* (2014, with Liou). Dr. Kapucu received his Ph.D. in Public and International Affairs from the Graduate School of Public and International Affairs of the University of Pittsburgh in 2003. He earned a Master of Public Policy and Management degree from Heinz College of Carnegie Mellon University in 1997. A detailed CV and other publications can be found at http://www.cohpa.ucf.edu/directory/naim-kapucu/.

List of Figures

List of Tables

Chapter 1
Introduction

Abstract In this chapter we discuss the goals and organization of the book and provide our perspectives and definitions of vulnerability, hazards, and resilience. We also describe, in detail, the data and methods utilized and conclude with a brief description of each chapter and the research gap the book is filling, particularly the analysis of the perception for the personnel responsible for disaster response and emergency management in Florida.

Keywords Disaster vulnerability • Hazards • Resilience • Focus groups • Survey data • Florida

Not a single day passes when we are not reminded of our vulnerability to disasters and other environmental hazards as the examples are abundant. The U.S. experienced several disasters, including 4 hurricanes in a span of 6 weeks in Florida in 2004, Hurricane Katrina in 2005, Super Storm Sandy in 2012, several deadly tornadoes in Alabama in 2011 and Oklahoma in 2013, among others. Worldwide there were several disasters including tsunamis in 2011 in Japan and Indonesia in 2004, alongside massive earthquakes in 2014 in Nicaragua and Chile. These events remind us disasters were, are, and will be a phenomenon experienced at some point in our life. These disasters, like many others, illustrate how disasters harm communities and individuals, and disrupt social-technical systems and community functions.

In response to these hazards and vulnerabilities, local, national, and international efforts have been established to identify and lessen disaster vulnerabilities by promoting the collaboration of different sectors in society, both public and private, in order to have disaster resilience communities. In the U.S., this emphasis is part of the "whole community" (Federal Emergency Management Agency 2011) approach to disaster management and recovery.

While the academic debates on the concept of resilience are plentiful (National Research Council 2010), limited studies have explored the perception for the personnel responsible for disaster emergency management and response. These are important to understand as they provide evidence on how perceptions sometimes can shape reality. Sometimes a reality might be separate from established policies and governance mechanisms. We argue that by analyzing the narratives from individuals representing different emergency management sectors, including county

© Springer International Publishing Switzerland 2015
F.I. Rivera, N. Kapucu, *Disaster Vulnerability, Hazards and Resilience*,
Environmental Hazards, DOI 10.1007/978-3-319-16453-3_1

emergency managers, nonprofit, and community groups, we can get a better grasp of the issues faced by those on the ground.

Although the focus of the book is on data from Florida, the implications can extend to other national and international contexts. The rationale for utilizing Florida as the anchor of the book is twofold. First, Florida is one of the United States' most at risk states for disasters. Weather patterns make many regions of the state vulnerable to hurricanes. Droughts often produce wildfires and unstable weather conditions are capable of producing tornados. Second, partly due to these constant disaster threats, the Florida emergency management system is ranked among the nation's best (Jordan 2006) and it has been identified as a model for the entire United States (Waugh 2006).

In addition to analyzing the perceptions of different sectors emergency management, the book explores the issues of disaster vulnerability, hazards, and resilience in rural community settings, an often neglected sector of study in the disaster literature. The capacity of rural communities to plan for and respond to disasters, in particular, has been the focus of recent studies (Cutter et al. 2010; Kapucu et al. 2013b; Skerratt 2013; Whitman et al. 2013). Compared to urban areas, rural communities may have a less diversified economic base and fewer financial resources to support disaster mitigation practices or rebuilding efforts. Thus, an emphasis of rural emergency management adds to this growing of the disaster research literature.

1.1   Key Terms: Hazard, Vulnerability, & Resilience

Before explaining the data sources and plan for the book, we want to provide our perspectives and definitions on the key concepts including hazards, vulnerability, and resilience. We define *hazards* as physical activities, phenomena, or human activities having the potential to cause injury, loss of life, damage to property, economic and social disruption, or environmental degradation (Kapucu and Özerdem 2013; McEntire 2004, 2005; Makoka and Kaplan 2005). Resilience is often defined as "a response to stress and can be considered as a theory that guides the understanding of stress response dynamics; a set of adaptive capacities that call attention to the resources that promote successful adaptation in the face of adversity; and a strategy for disaster readiness against unpredictable and difficult to prepare for dangers (National Research Council 2009, p. 23)." We define *resilience* as "the ability to adapt trough the redevelopment of the community in ways that reflect the community's values, and goals, and its evolving understanding of external forces with which it must contend" (Kapucu et al. 2013a, b, p. 357). This definition takes into account resilience is more about the ability of a community or system to "bounce back," but rather a process by which communities confront and try to resolve different social, political, and economic forces impacting the way they prepare, mitigate, response and recover from a disaster.

In the context of the book, we draw from the work of Kusenbach and Christmann (2013) to define *disaster vulnerability* as "a concept that denotes a social practice in which a certain unit (a subject, a group, or any kind of system) is placed at the center of a complex analysis of injury" (p. 64). We argue to truly understand vulnerability attention must be paid to important issues such as: the social construction of disaster vulnerability including the cultural context of human perceptions and interpretations (particularly the perceptions of those in the front lines of disaster emergency management and response); the view of vulnerability as negative and resources as positive while often overlooking the unintended side effects of vulnerabilities and resources; and the complexities of time which alter the meaning of vulnerabilities before, during, and after a disaster situation.

1.2 Data Sources

1.2.1 Survey Data

We collected data through a mail and online survey of emergency management professionals in eight Central Florida counties. Central Florida houses approximately 2.2 million people, which is nearly 12 % of the total population of the state (Metro Orlando Economic Development Commission n.d.). The purpose of the survey was to identify the communication, coordination, and resource sharing procedures among organizations involved in emergency management within Central Florida. The survey instrument also included questions on mitigation strategies, planning, preparedness, response, recovery, partnerships, organizational capacity, and demographic information. Results for close-ended questions were in the form of a 5-item Likert scale (e.g. 1 being strongly disagree to 5 being strongly agree). After the survey questionnaire was prepared, it was sent out for review to a panel of experts consisting of 25 organizations representing each county (2–3 organizations per county). The panel of experts included emergency managers of study counties and organizations that emergency managers recommended to include in the panel. The survey was subsequently revised based on the panel recommendations (see Appendix A). County emergency managers assisted in administration of the survey except three counties (Lake, Flagler, and Levy), as Emergency Managers of these counties did not want to participate in this survey. Emergency managers of the remaining eight counties shared the survey with the organizations they listed in their Comprehensive Emergency Management Plan (CEMP) as having either a primary or secondary role in disaster response. The administration of the survey started on August 19, 2011 and ended on January 20, 2012. This time frame covers approximately 3 months before and after the Atlantic hurricane season. In total, 242 organizations responded to the survey, which accounts for a 38.0 % response rate. After eliminating responses in surveys with too many missing variables, the useful response rate fell to 25.2 % (Kapucu et al. 2013a).

1.2.2   Focus Group Data

A series of semi-structured focus groups were conducted between November 2011 and March 2012, as part of a larger project analyzing rural disaster resiliency in Central Florida. In all, seven focus groups were conducted in the following Central Florida counties: Brevard, Lake, Orange, Osceola, Seminole, Sumter, and Volusia. Focus group participants included citizen groups, nonprofit organizations, faith-based community organizations, emergency management agencies, and business representatives. A total of 60 individuals, representing 20 unique organizations, participated in the focus groups. Recruitment of participants consisted of contacting organizations listed in the county comprehensive emergency management plan developed by each county and are accessible by the general public. For each county we asked emergency management organizations and community representatives from nine emergency support functions: transportation, infrastructure, first responders, information, health care, support, food and water, utilities, and communications to participate in the focus group. An interview guide (See Appendix B) was designed to explore themes related to hazard mitigation practices, emergency response networks, disaster recovery methods, community resiliency, and networks used by rural counties and rural parts of urban counties in Central Florida. Specifically, the interview script included the following themes: mitigation/preparedness/response/recovery, community vulnerability and disaster resilience, community relations and adaptation, social media/news media, politics and government action, and special need populations. The number of participants for each focus group ranged from 5 to 13. The interviews were conducted at an agreed upon location convenient to the participants and lasted on average between 1 and 2 h. Two faculty members and three graduate research assistants facilitated the focus groups.

   Before data collection data began, the university institutional review board (IRB) approved the study instrument and research protocol. Interviews were conducted in English, except for the Volusia County focus group, which was conducted in Spanish. A letter of consent was provided to each participant prior to the start of the focus group meeting and interviews were digitally recorded. At the conclusion of the project, the interviews were transcribed, verbatim, by a member of the research team, resulting in approximately 140 pages of data.

1.3   Data Analyses

For the survey data, we analyzed the open-ended questions included in the questionnaire. Respondents were asked the following questions: How do you define disaster resilience? Are there additional elements (not covered in this survey) you think are important to create disaster resilient communities? And what are the obstacles to build disaster resilient communities? The responses to these questions were analyzed utilizing a specialized software program (SPSS/STATA). The results for the survey data is analyzed and discussed in Chap. 5.

The analyses and coding for this focus group was derived from a concept driven perspective know as framework analysis (Gibbs 2007; Ritchie and Lewis 2003). A list of thematic ideas drawn from the disaster resiliency literature was built. Afterwards, members of the research team read the transcripts and assigned sentences to different thematic domains based on the interview script. At the conclusion of this process, a meeting took place to review the themes until consensus was reached.

Research participants were identified as respondents with no identifiable traits outside of gender. Based on the interview script, the concept-driven themes, and the research team consensus, participant quotes were divided into several themes that are discussed throughout the book including: hazards (Chap. 3), vulnerability (Chap. 4), resilience (Chap. 5), the path to resilience (Chap. 6), rural communities (Chap. 7) and farmworkers (Chap. 8).

1.4   Plan for the Book

This book offers a unique examination into the perceptions of disaster vulnerability and resilience and will contribute to the literature on emergency management, policy and planning as it pushes the resilience and vulnerability debate from the theoretical to the practical realm incorporating perceptions of what is currently happening in disaster management groups and organizations. It will also contribute to the sociological literature as it analyzes the disaster emergency personnel perspectives on vulnerable groups, such as the elderly, children, and farmworkers, among others. Ultimately, this book sheds new light on how disaster emergency personnel perceive disaster resilience and vulnerability and provides evidence on how perceptions sometimes can shape reality. A reality separate from established policies and governance mechanisms.

In Chap. 2 we provide a comprehensive overview of the emergency management system, structure, policies, and perspective in Florida. The system contains important intergovernmental relations and organizational features that will be explained in detail. Although there are no comprehensive updated studies on emergency management system and practice in Florida, many scholars highlight the effectiveness of this particular system (Waugh 2006). This chapter will also include federal level policies and administrative structure to the chapter, as they are integral part of the emergency management system.

In Chap. 3, we detail an overview of some of the natural disaster hazards in Florida (hurricanes, wildfires, tornadoes, among others) and the measures taken by emergency managers to prevent/mitigate, prepare, response, and recover from them (Kapucu et al. 2008). We do so by providing a content analysis of news articles after the 2014 hurricane season with a particular focus on the rural areas of our study area. In Chap. 4, we discuss the concept of vulnerability with a detailed analysis of the social, economic, geographical, and political attributes of Florida communities and regions that shape vulnerabilities to disasters. The dialogue is based on the results from the survey and focus groups data and the participant's identification of perceived vulnerabilities including poverty, homelessness, the elderly, and others. In addition, we examine several disaster vulnerability identification tools and disaster risk/vulnerability reduction strategies.

Chapter 5 extends the previous chapter discussion of vulnerability by analyzing the conceptualizing of resilience by the study participants, including a detailed analysis of corresponding conceptualizations including: bouncing back, restoring, avoidance, and others. We pay particular attention to differences between urban and rural settings along with established disaster policies and tools to assess the vulnerability of communities.

In Chap. 6, we review the obstacles to resilience identified in the research literature and discuss the adaptive resilience and community capital framework. We build on the frameworks and provide a detailed analysis of the study participant's responses with regards to what they perceived to be obstacles to resilience, particularly issues relating to apathy and complacency, communication issues, and funding, among others (Wang and Kapucu 2008). We also discuss differences in the responses by urban and rural settings.

Chapter 7 addresses how rural communities and their residents respond to natural and man-made hazards (Brennan and Flint 2007; Bankoff et al. 2004; Pelling 2003). Compared to urban areas, rural communities may have a less diversified economic base and fewer financial resources to support disaster mitigation practices or rebuilding efforts. Moreover, the separation and remoteness of rural communities from urban areas, low population density, and inadequate communication networks pose challenges particular to rural communities (Janssen 2006; Oxfam America 2009). In this chapter we use the adaptive resilience and community capital framework to discuss the findings from focus group data particularly, local policies and support, community capital, capital vulnerability, and private support previously identified as key to disaster resilience, adaptive resilience and adaptive learning.

For Chap. 8, we discuss our focus group results with a group of migrant farmworkers. This group has received some attention in the research literature particularly on their vulnerabilities, such as lack of access to health services (Carrion et al. 2011; Castañeda et al. 2010), a lack of social integration (Bail et al. 2012; Lichter 2012), substandard housing conditions (Ziebarth 2006), and lack of trust of government (Chavez et al. 2006) among others. Utilizing data from a focus group with farmworkers in this chapter, we discuss the unique challenges to disaster resilience being faced. In addition, we discuss some of the collective action measures taken by this group, non-profits, and local emergency managers to increase their disaster resilience.

Finally, in Chap. 9, we examine the policy, practical, and theoretical implications of the research. First, we provide a much needed and updated description of the Florida disaster management system, lauded as a model for the nation. Second, we provide first-hand accounts of county emergency managers, nonprofit, and community groups in relation to different issues including vulnerability and resilience. Third, we discuss perceived differences to disaster resilience between urban and rural settings. Fourth, we give voice to the unique challenges to disaster resilience experienced in rural communities and non-represented groups, such as farmworkers. Finally, this chapter concludes with how the findings bring new evidence and insights to the study of disaster resilience, how it integrates knowledge from sociology and public policy and governance to the study of natural disasters, and how the book provides useful and accessible insights not only to academic circles, but to readers in government, nongovernmental organizations, and the private sector.

References

Bail, K. M., Foster, J., Dalmida, S. G. et al. (2012). The impact of invisibility on the health of migrant farmworkers in the southeastern United States: A case study from Georgia. *Nursing Research and Practice, 2012*, 8 pages. doi:10.1155/2012/760418.

Bankoff, G., Frerks, G., & Hilhorst, D. (2004). *Mapping vulnerability: Disasters, development and people*. London: Earthscan.

Brennan, M. A., & Flint, C. G. (2007). Uncovering the hidden dimensions of rural disaster mitigation: Capacity building through community emergency response teams. *Southern Rural Sociology, 22*(2), 111–126.

Carrion, I. V., Castañeda, H., Martinez-Tyson, D., & Kline, N. (2011). Barriers impeding access to primary oral health care among farmworker families in Central Florida. *Social Work in Health Care, 50*(10), 828–844.

Castañeda, H., Carrion, I. V., Kline, N., & Martinez Tyson, D. (2010). False hope: Effects of social class and health policy on oral health inequalities for migrant farmworker families. *Social Science & Medicine, 71*(11), 2028–2037.

Chavez, M. L., Wampler, B., & Burkhart, R. E. (2006). Left out: Trust and social capital among migrant seasonal farmworkers. *Social Science Quarterly, 87*(5), 1012–1029.

Cutter, S. L., Burton, C. G., & Emrich, C. T. (2010). Disaster resilience indicators for benchmarking baseline conditions. *Journal of Homeland Security and Emergency Management, 7*(1), 1–24.

Federal Emergency Management Agency. (2011). *A whole community approach to emergency management: Principles, themes, and pathways for action*. Washington, DC: U.S. Department of Homeland Security, Federal Emergency Management Agency.

Gibbs, G. (2007). *Analyzing qualitative data*. Thousand Oaks: Sage.

Janssen, D. (2006). Disaster planning in rural America. *Public Manager, 35*(3), 40–43.

Jordan, L. J. (2006, February 15). *Report slams Katrina disaster preparation*. Associated Press Wire Service.

Kapucu, N., & Özerdem, A. (2013). *Managing emergencies and crises*. Boston: Jones & Bartlett Publishers.

Kapucu, N., Berman, E., & Wang, S. (2008). Emergency information management and public disaster preparedness: Lessons from the 2004 Florida Hurricane season. *International Journal of Mass Emergencies and Disasters, 26*(3), 169–197.

Kapucu, N., Hawkins, C., & Rivera, F. (2013a). *Disaster resilience: Interdisciplinary perspectives*. New York: Routledge.

Kapucu, N., Hawkins, C. V., & Rivera, F. I. (2013b). Disaster preparedness and resilience for rural communities. *Risk, Hazards & Crisis in Public Policy, 4*(4), 215–233.

Kusenbach, M., & Christmann, G. (2013). Chapter 4: Understanding hurricane vulnerability. In N. Kapucu, C. Hawkins, & F. Rivera (Eds.), *Disaster resiliency: Interdisciplinary perspectives* (p. 61–83). New York: Routledge.

Lichter, D. T. (2012). Immigration and the new racial diversity in rural America. *Rural Sociology, 77*, 3–35.

Makoka, D., & Kaplan, M. (2005). *Poverty and vulnerability. An interdisciplinary approach.* Munich: Universitat Bonn.

McEntire, D. A. (2004). Tenets of vulnerability: Assessing a fundamental disaster concept. *Journal of Emergency Management, 2*(2), 23–29.

McEntire, D. A. (2005). Why vulnerability matters: Illustrating the need for a modified disaster reduction concept. *Disaster Prevention and Management, 14*(2), 206–222.

Metro Orlando Economic Development Commission (MOEDC). (n.d.). *Diversity, Orlando, MSA.* http://www.orlandoedc.com/core/fileparse.php/98857/urlt/Demographics_Diversity.pdf. Accessed on 1 Dec 2013.

National Research Council (NRC). (2009). *Applications of social network analysis for building community disaster resilience.* Washington, DC: The National Academies Press.

National Research Council (NRC). (2010). *Building community disaster resilience through private-public collaboration.* Washington, DC: The National Academies Press.

Oxfam. (2009). *Exposed: Social vulnerability and climate change in the US Southeast.* Washington, DC: Oxfam.

Pelling, M. (2003). *The vulnerability of cities: Natural disasters and social resilience.* London: Earthscan.

Ritchie, J., & Lewis, J. (2003). *Qualitative research practice: A guide for social science students and researchers.* Thousand Oaks: Sage.

Skerratt, S. (2013). Enhancing the analysis of rural community resilience: Evidence from community land ownership. *Journal of Rural Studies, 31*, 36–46.

Wang, X., & Kapucu, N. (2008). Public complacency under repeated emergency threats: Some empirical evidence. *Journal of Public Administration Research and Theory, 18*(1), 57–78.

Waugh, W. L., Jr. (Ed.). (2006). Shelter from the storm: Repairing the national emergency management system after Hurricane Katrina. Special issue of *The Annals of the American Academy of Political and Social Science, 604*, 256–272.

Whitman, Z. R., Wilson, T. M., Seville, E., Vargo, J., Stevenson, J. R., Kachali, H., & Cole, J. (2013). Rural organizational impacts, mitigation strategies, and resilience to the 2010 Darfield earthquake, New Zealand. *Natural Hazards, 69*, 1849–1875.

Ziebarth, A. (2006). Housing seasonal workers for the Minnesota processed vegetable industry. *Rural Sociology, 71*, 335–57.

Chapter 2
Emergency Management in Florida

Abstract This chapter provides an overview of the emergency management system in the State of Florida. The Florida emergency management system contains important intergovernmental relations and organizational features that will explain in detail. Many scholars highlight the effectiveness of Florida's emergency management system. Yet, there are no comprehensive updated studies on emergency management system and practice in Florida. This chapter provides a comprehensive system, structure, policies, and perspective of emergency management in Florida. The chapter also includes federal level policies and administrative structure to the chapter as they are integral part of the emergency management system in the US. We conclude that for a community to increase their capacity for resilience, they must prioritize emergency management and hazard planning policies while also creating supportive administrative structures. Furthermore, transparency is necessary to help eliminate the numbness generated by underestimation and under-preparedness.

Keywords Emergency management • Disaster resilience • Disaster preparedness • Disaster planning • Disaster response • Rural communities • Florida

2.1 Emergency Management in the US: Intergovernmental Perspectives

When it comes to the emergency management in Florida, there have been a number of changes due to the negative effects of several natural disasters. The field has been considered 'broken' with government officials and organizations wanting to know the reasons behind disconnects and how to correct them (Kapucu and Özerdem 2013; Mittler 1997; Rubin 2013). An understanding came about for emergency managers needing to be knowledgeable of the past, current and potential future policy processes and to comprehend the existing intergovernmental system, which exists as an integration of emergency management phases into a challenge-filled

© Springer International Publishing Switzerland 2015   9
F.I. Rivera, N. Kapucu, *Disaster Vulnerability, Hazards and Resilience*,
Environmental Hazards, DOI 10.1007/978-3-319-16453-3_2

structural framework (Mushkatel and Weshcler 1985). For successful implementation and policy generation, Mittler (1997) detailed the following needs:

1. The existence of a widely recognized problem that the emergency management system was inadequate to serve the needs of the state in the event of major natural and other disasters;
2. The support of the governor, many legislators, and the emergency management professionals in the state for comprehensive change and a dedicated source of funding;
3. The long-term development of a program of change, which had fostered previous legislation, thereby establishing a foundation for the drafting of a new bill;
4. The use of a funding mechanism, which did not increase, taxes or divert general revenue funds from other programs (p. 12).

With some natural disasters requiring federal response, it is imperative for government officials to know where responsibilities lie. For example, Hurricane Andrew of 1992 was an unprecedented event needing response from the Federal Emergency Management Agency (FEMA). However, critiques felt FEMA's response was a weak link due to the organization's head, Wallace E. Stickney, who had no previous experience in emergency management (Hughes 2012; Rubin 2013).

When communicating about disasters, the definition of the event can be difficult to understand due to its encompassing nature. Wilson and Oyola-Yemaiel (2001) pose the meaning of 'disaster' to include "human responses and adaptations to events…the social structure is disrupted and the fulfillment of all or some of the essential functions of society is prevented" (p. 118). Speaking of these events, emergency management is seen as "the discipline and profession of applying science, technology, planning, and management to deal" with disruptions of community life (p. 118–19).

When defining resilience, one must take into account the conditions of the disaster effecting individuals and communities (Kapucu 2012b). Due to the importance of context, adaptive capacity embraces degrees of cultural differences between local officials, planners, emergency managers and more, to increase understanding and empower citizens. The four components of adaptive capacity include: social capital, community competence, information and communication, and strong economy. If a community increases their ability to coordinate response efforts, then they decrease their vulnerability, or sensitivity of their system (Kapucu 2012b; Sylves 2008; Waugh 2006).

Commonly created frameworks of disaster response usually incorporate the following four elements: mitigation, preparedness, response and recovery with each area being intrinsically linked with the next. Starting with mitigation, emergency management personnel begin to act in preventing disaster and attempting to reduce, or lessen, the damage. Preparedness incorporates the addition of planning processes and organizational relationships to improve a community's capacity of response. Even though disasters are unpredictable, first responders usually include local officials who "collectively interpret and make sense of their environment" (Kapucu 2008, p. 245). Lastly, the recovery phase includes dealing with the aftermath.

2.2 Emergency Management in Florida

National Academy of Public Administration (NAPA) (1993), in their report after hurricane Andrew, recommends the state and localities develop an effective emergency management system in response to and recover from emergencies and small-scale disasters. Federal government can be part of response and recovery in case of catastrophic or major disasters. Emergency management system in Florida represents some similar characteristics with the U.S. system in general. The Division of Emergency Management (DEM) is a state level counterpart of FEMA at the national level. Each county has an emergency management office as well. Emergency management professionals in Florida are not only dealing with unique structures at the city and county levels, but they are dealing with the increase in population along with the diversity brought by environmental factors. The community vulnerability has increased resulting in a higher risk to natural disaster exposure (Choi 2004).

Disaster resilience is crucial in Florida due to its complex historical context, cultural and sociological factors, agriculture, economic factors, insurance industry and tourism (Kapucu 2012b). In addition, Florida is one of the fastest growing states in terms of population, which adds stress to response and recovery efforts (Wilson and Oyola-Yemaiel 2001). In layman's terms, resilience is the ability to recover from a disruptive change. The issue is not necessarily if managers are asking the right questions, but are they asking the questions correctly given context (Collier 2012). More importantly, emergency managers are in the position to create a common language and role clarification for disaster response (Wilson and Oyola-Yemaiel 2001) (Fig. 2.1).

2.2.1 Major Disasters

To understand the development of emergency management in Florida, one must become familiar with focusing events (Birkland 2007), or trigger events, distinguished based on their level of damage and categorized as a crisis, disaster or catastrophe. Events such as Hurricane Andrew in 1992 and the season of 2004, when Hurricanes Charley, Frances, and Jeanne swept through, were trigger events resulting in necessary policy changes at "the county level in hazard mitigation practices" due to their effect on the area of disaster response (Hawkins and Knox 2014, p. 111). For instance, the response of local officials can negatively or positively affect how Emergency Support Functions (ESFs) operate. The primary agency of the ESF, although supported by other organizations, is responsible for the "mitigation, planning, protection, response, and recovery from hazards and emergencies…based on its expertise, authority, resources, and capabilities" (p. 118) (Fig. 2.2).

Similar to at the national level, emergency management responsibilities and agencies functioning that role are organized around ESFs. "During times when disasters are coordinated between different of levels of government the ESF-based structure is specifically important. The standardization of resource grouping as well

Fig. 2.1 Florida's 67 counties (MapWise 2013)

as the responsibilities of actors leads to a more streamlined response and recovery process" (Kapucu 2012a, p. S44). When forecasting, a hurricane has the ability to change or alter their path. Climate change is an area predicted to increase the chance of super storms (Kerjen 2008). If scientists are already projecting highly impactful hurricane seasons then it is up to emergency personnel to model action verses complacency. In addition, managers need to learn from disastrous events like scientists who use modeling to determine potentially devastating situations.

2.2.2   Hurricane Andrew

When determining the beginning to the Florida emergency management era, some researchers denote the response to Hurricane Andrew as the event which started it all (Hall 2011). Hurricane Andrew is considered the costliest storm in U.S. history.

ESF Number	Federal	Florida
1	Transportation	Transportation
2	Communications	Communications
3	Public works and engineering	Public works
4	Firefighting	Firefighting
5	Emergency management	Info and planning
6	Mass care, emergency assistance, housing, and human services	Mass care
7	Logistics management and resource support	Unified logistics
8	Public health and medical services	Health and medical
9	Search and rescue	Search and rescue
10	Oil and hazardous materials response	Hazmat
11	Agriculture and natural resources	Food and water
12	Energy	Energy
13	Public safety and security	Military support
14	Long-term community recovery	Public information
15	External affairs	Volunteers and donations
16		Law enforcement
17		Animal services
18		Business, industry and economic stabilization
19		Damage assessment (Orange County)
20		Utilities (Orange County)

Fig. 2.2 Emergency Support Functions on the federal level and for Florida (Hawkins and Knox 2014)

It was the fourth tropical cyclone in the 1992 season and was unprecedented in its intensification and ultimate damage. The storm caused a spotlight to occur on the complex administrative and political system in place, which allowed, or caused, the inadequate response (Wamsley and Shroeder 1996) as Florida incurred $30 billion in damages and 52 losses of life (Hawkins and Knox 2014). Ever since 1992, due to Hurricane Andrew, the legislature has maintained an Emergency Management Preparedness and Assistance Trust Fund to assist with property insurance issues and, hopefully, eliminate the financial burden to local governments (Choi 2004).

Other important legislative effects came into play with the creation of the Emergency Management Association Compact (EMAC) created by Florida governor Lawton Chiles (Kapucu et al. 2009). Governor Chiles was disappointed with the response efforts of Hurricane Andrew and advocated for a mutual aid agreement between states when natural disasters have magnanimous negative effects. The response and recovery efforts could now focus more on formalized partnerships, along with informal, and contribute towards positive mitigation without needing federal approval. Moreover, these collaborations allow for trust building, respect and continual interactions, which can ultimately enhance a community's capacity (Kapucu et al. 2009).

Although Hurricane Andrew caused major devastation to local communities, it did not hurt local morale in a permanent manner. Twigg (2012) speaks to natural disasters and their affinity to generate leadership in areas outside the political arena. "The core value of people after a disaster is helping victims; a second tier of values includes maintain public morale (Dynes 1970). People in a disaster area are frequently optimistic about rebuilding and the future of their cities" (Twigg 2012, p. 162).

2.2.3 Four Hurricanes in 2004

Although local governments felt they were more prepared for natural disasters, such as hurricanes, the four storms in 2004 were more than they predicted. A succession of three storms tested emergency management policies at that point. Beginning with Hurricane Charley in August of 2004, followed by Hurricane Frances in September, and ending with Hurricane Jeanne at the end of September, emergency managers found themselves scrambling. Hurricane Jeanne powered through central Florida stretching already tight staffing reserves and highlighted an issue of emphasizing basic needs of Emergency Operations Center (EOC) staff and ESF training (Hawkins and Knox 2014).

The aftermath of the 2004 hurricane season was a time of reflection, in terms of response and planning, for emergency personnel as they totaled up damages ranging by county from $8 to $16 billion (Hawkins and Knox 2014). What can they do to more adequately manage an event? By posing critical questions, officials can improve their practices. Therefore, researchers propose mandating after action reports to evaluate and facilitate needed policy changes and/or improvements (Hawkins and Knox 2014) (Fig. 2.3).

2.2.4 Key Policies

"Florida began to address coastal management, disaster preparedness, and hazard mitigation systematically during the 1970s in response to a growing awareness that the state was highly vulnerable to coastal storms" (Mittler 1997, p. 6). Due to the lack of response for declared emergencies, local governments have encountered over-response and delay during the management process (Choi 2004). "First, information on the determinants of emergency management growth can contribute to sound fiscal policies. Second, recognizing the determinants of emergency management growth can help decision makers design local emergency management strategies or plans. Third, county policy makers can effectively deal with problems that are related to emergency management if they know the factors affecting emergency management growth" (Choi 2004, p. 212). However, none of these positive moves were being done. In fact, a critique on the field of emergency management has been the past disconnect between practitioners and academicians. "While practitioners

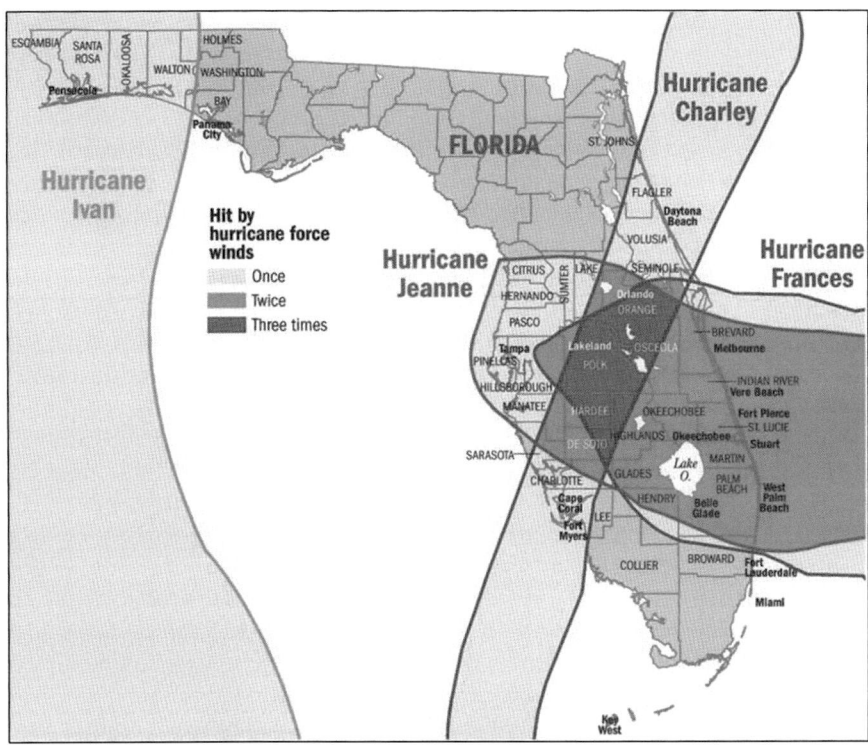

Fig. 2.3 Pathways of the hurricanes of 2004 (Kapucu 2008)

have long known how political a major disaster may become, the academic community has lagged in that appreciation. Only a relative handful of political science scholars have focused on what the disaster research com-munity calls low-probability high-consequence events" (Olson and Gawronski 2010, p. 206).

Moreover, along with other factors increasing a community's risk during a disaster, the political ideology is stated to have a substantial effect on emergency management growth (Choi 2004). Emergency management has a long-standing legislation beginning in 1803 with the passing of a resolution allowing aid to multiple individuals. Policies continued to develop in 1974 with the Disaster Relief Act, amended in 1988 to become the Disaster Relief and Emergency Assistance Act (Hughes 2012). Critical legislation, however, was created after catastrophic events such as the attacks on the World Trade Center in 2001 where President George W. Bush issued the Homeland Security Presidential Directive 5 shortly followed by the National Incident Management System (NIMS), which included the key component of Incident Command Systems (ICS) (Hughes 2012).

Along with pivotal legislation, response frameworks were generated after events, like Hurricane Andrew, to outline state and local efforts of alleviating impact. The National Response Plan (NRP) was formulated to "align federal coordination structures, capabilities, and resources into a unified, all-discipline, and all-hazards approach to domestic incident management" (Hughes 2012, p. 4). However, these plans are only as effective as the managers and citizens allow them to be. Once a disaster occurs, there is a critical time period where morale is low. "Public estimation of government is especially volatile in post-disaster situations because (i) a substantial portion of the population has been suddenly and visibly reduced to the search for the most elemental material needs;(ii) the media are covering the disaster and then the response with unusual and sustained intensity, essentially putting all aspects of the losses and of the response under a public microscope; and (iii) the general public is unusually attentive, at least for a time" (Olson and Gawronski 2010, p. 208). Ironically, events where administrators acted in positive ways and were not mentioned in the media channels to the public at large (Kapucu and Van Wart 2008) (Fig. 2.4).

2.2.5 Organizations

Comparing disaster response, researchers noted an increased need to focus on improving "communications across agencies that plan for and respond to emergencies" (Hawkins and Knox 2014, p. 121). Disaster response is a "complex interaction among multiple government agencies, non-profit organizations, private businesses, and individual citizens" (Kapucu 2006, p. 256). The Federal Emergency Management Agency (FEMA), created in 1979, is a major collaborative entity in disaster response

Fig. 2.4 Orange County, FL, Emergency Operations Center (Source: Authors)

as the agency assists in the response to "administrative and structural difficulties" and essential functions while emphasizing a comprehensive management approach (Wilson and Oyola-Yemaiel 2001, p. 119). Furthermore, FEMA provides a unique connection to local officials by ensuring federal monies are being utilized in a useful way, consistent with federal policy, and facilities support to increase emergency management capability.

2.2.6 Financing Disaster Preparedness in Florida

Florida presents a committed government support to disaster preparedness through designated tax revenue. "Without having the opportunity to raise funds through personal income taxes and also having a strong predilection against increasing general revenue obligations or issuing general obligation bonds, three of the most common sources of state revenue, Florida has been forced to rely on revenue bonds (which identify explicit revenue to repay borrowing and do not require voter confirmation) and user taxes as its primary means of public financing or it has consented to increased local taxation and bonding for specific purposes" (Mittler 1997, p. 4).

When debating on a financial emergency being triggered, the following determining factors must be met: (a) failure to repay short term loans or make debt service payments, (b) inability to transfer at the proper time, (c) failure to pay for one pay period, (d) procure two successive years of deficits for which the government is incapable of covering, and (e) lack of compliance with local government retirement system as stipulated by law (Adams 1997, 2014). "The state's role, rather, is to insure that local governments take the appropriate steps to eliminate the financial emergency. It is difficult for the state to step in before problems become emergencies and not seriously erode home-rule authority" (Adams 1997, p. 1).

To increase a county's ability to sustain emergency management growth, there are several propositions to take into account:

- The greater the county economic development, the greater the emergency management growth.
- Relationships between citizen political ideology and local emergency management growth will exist.
- Counties with council elected executive forms will support greater emergency management growth.
- The higher the community vulnerability to disaster, the greater the emergency management growth.
- The higher the population density in a county and the greater the rate of population growth, the greater the emergency management growth.
- The greater the racial homogeneity within a county, the greater the emergency management growth (Choi 2004, p. 215).

To reduce the impact of the recent economic crisis "The Government Accountability Office projects that to balance their budgets for the long term, state and local governments would need to reduce spending or increase taxes by an amount equivalent to approximately one-eighth of their current expenditures" (FEMA 2011, p. 3). With the financial state of affairs, it is important to raise money for disaster response in a way that does not negatively impact the citizens of Florida's communities. An example of engaging the public in preparation is with the 2014 tax cut legislation passed by Governor Rick Scott. Mayor Buddy Dyer, from the City of Orlando, supported the tax due to resident's inability to properly prepare for disasters like hurricanes due to a sense of complacency (Scott 2014).

To address complacency, Kapucu (2008) promoted community coordination in the aspect of combating unawareness, providing consistent and clear information, being aware of previous past warnings, and encouraging citizens to implement recommended responses. Coordination can be especially useful in the rural areas of Florida where informing citizens can take more of an effort due to limited communication tools (Kapucu et al. 2008, 2013a, b) (Fig. 2.5).

An example of positive strides to eliminating harsh financial debt, due to natural disasters, is the Florida Emergency Management Preparedness and Assistance Trust Fund. The fund is created of mutual aid agreements between local government officials and the state government and is influenced by certain insurance policies to generate a bank of money during disaster situations (Legislative Committee on Intergovernmental Relations 2007). The monies allow for emergency operations planning by aligning the goals of both parties (Hawkins and Knox 2014; Wilson 1998). The reality is financial capacities are higher at the federal level verses the state and local levels, which perpetuates the need for collaboration (Waugh 1994).

For instance, "Agranoff and McGuire (2003) explained that collaboration can occur on both vertical and horizontal dimensions. Vertical collaboration emphasizes work across levels of governments within the U.S. federal system, while horizontal collaboration refers to joint works by jurisdictions on the same level of government. In the case of local emergency management and homeland security, flexible collaboration in both vertical and horizontal contexts are necessary" (Chang 2012, p. 15). Furthermore, Stalling and Quarantelli (1985) argues that citizen behavior and emergent groups are not limited to individual efforts in disaster response. In metropolitan settings civic engagement and citizen initiatives have an important role in multiple aspects of managing Kapucu (2012a) argues that "they are not only

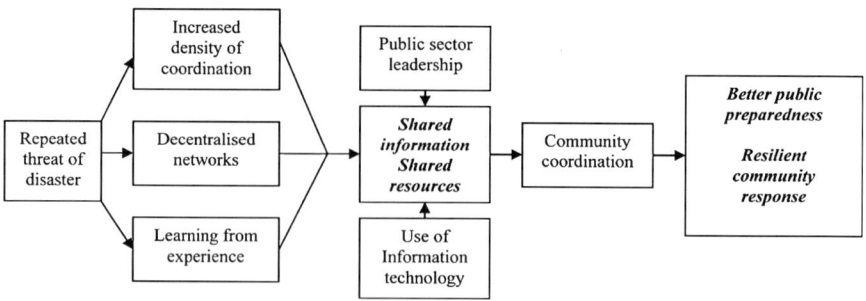

Fig. 2.5 Example of community coordination (Kapucu 2008)

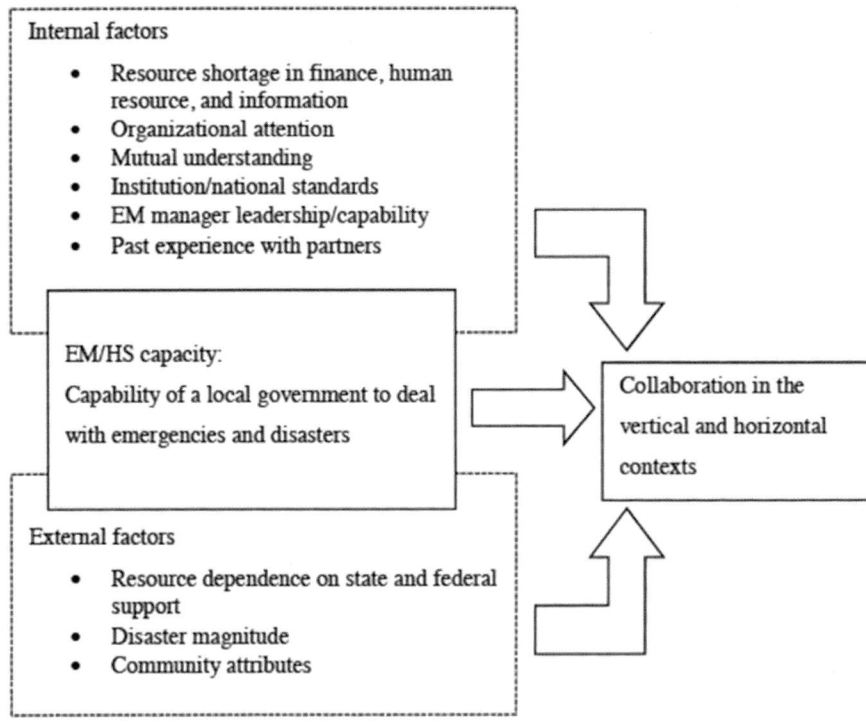

Fig. 2.6 Theoretical framework for vertical and horizontal collaboration (Chang 2012)

involved in preparedness and response efforts, but also provide cash support and donate blood in the aftermath of catastrophes" (p. S42) (Fig. 2.6).

Careful consideration must be given when declaring a financial issue in a disaster situation. Local governments need to consider four related areas: early warning, notification, definition, and roles and responsibilities (Adams 1997). With many disasters having unforeseen consequences, it can soon become unclear as to whose role it is to response and in what way. Therefore, understanding each organization's power and resources is important (Kerjen 2008).

2.2.7 All Hazards Approaches for Disaster Resilience

Since the field of emergency management has historically been collaborative, starting in the twentieth century, there has been the prevalence of multi-sector collaboration (Kapucu 2012a). These collaborations are imperative to creating a comprehensive, multi-hazard disaster plan based on actual needs. The focus is intertwined with response and recovery where personnel attempt to improve methods of communication, flow of information, coordination efforts, and relations with authorities (Stripling 2013). With mitigation being a crucial component in disaster response and resiliency, planning is highly important. "'Our local mitigation strategy has become a priority document for the county government, and I know that its implementation will help us create a disaster-resistant future for all of our citizens and visitors,' says Richard Crotty, Orange County chairman" (American City and County 2001, p. 13).

A key to mitigation policy is the aspect of development, which "is conceptualized as the production of our built environment" (Hawkins and Knox 2014, p. 112). Areas to view this progress are through the responsibilities of county governments during emergency management activities where local officials coordinate municipal activities within their regions or jurisdictions. "Mitigation strategies have become embedded in local land-use plans, zoning ordinances, building codes, and local education programs" (Brody et al. 2009, p. 913).

To lead the emergency management system in a positive manner, administrators are recommended to adopt a Comprehensive Emergency Management outlook where the four component framework is used in tandem with federal, state, and local government role clarification to coordinate services and disaster efforts (Mittler 1997). An important characteristic of CEM is the knowledge of the intergovernmental system and the investigation into how the context of the disaster affects policies and procedures to increase coherency (Mushkatel and Weshcler 1985) (Fig. 2.7).

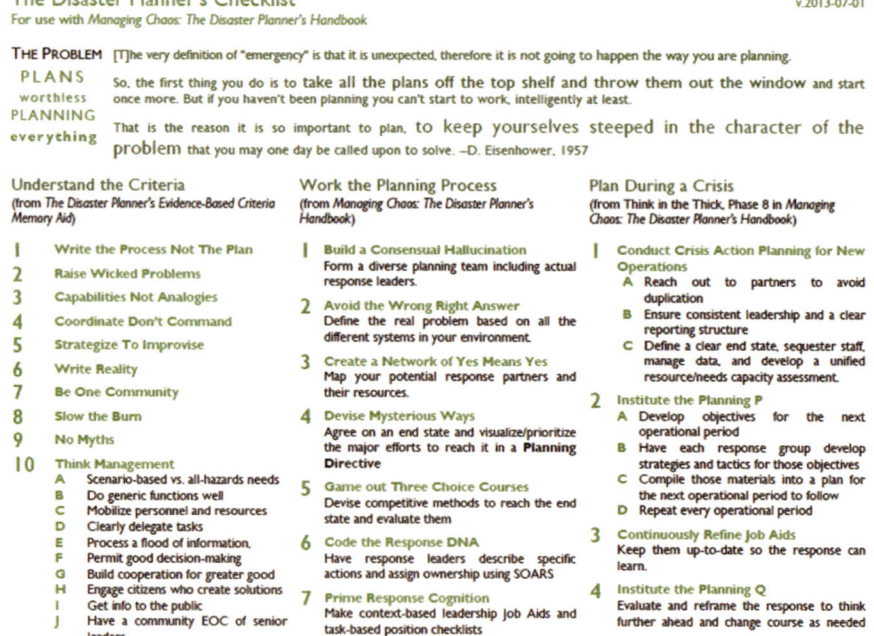

Fig. 2.7 Disaster planning criteria to help create comprehensive plans (Stripling 2013)

Furthermore, the current structure is focused on the cooperation between levels of government verses the primary responsibility falling to one organization or agency. "Where states follow the comprehensive planning approach, achieving state goals depends on two tiers of implementation: (1) state agencies implementing legislated state goals and objectives and (2) local governments implementing state agency directives" (Deyle et al. 2008, p. 350).

Due to the need for a reciprocal relationship, it is imperative for local government officials to work positively with state representatives. The capacity for a community to rebound from a disaster depends on a mutually beneficial situation. "Lindell and Meier (1994) for example, have illustrated that community support, vulnerability to hazards, staff and structures, and emergency planning resources are considered important variables to make emergency management effective" (Choi 2004, p. 213). In most cases, the disaster will require more resources than any one organization has, which means intergovernmental cooperation is key (Waugh 1994). "Developing trust among agencies is done by mutual learning and action in a network. The actions that most build trust are the completion of accepted assignments, follow-through, and commitment to the cause" (Kapucu et al. 2009, p. 300).

Within a CEM plan, there is an emphasis on development and education. One avenue to gain this insight is through the Governor's Hurricane Conference, initiated in 1987, which holds workshops focused on all phases of emergency response and is supported by connected organizations (Mittler 1997). Key questions consist of how to: "develop sophisticated and successful actions or strategies in order to alleviate the negative effects of a disaster; reduce the vulnerability and risk factors affecting communities, cities, industries, businesses, and people; and sustain the ability to recover faster" (Kapucu 2012b, p. 205).

With environmental factors being an underlying cause of natural disasters, Florida is stated to be one of the most prone states for emergency situations due to the expansive coastal regions (Rosenfeld 2006). The importance lies in having the conversation of how the environment could predict or cause future natural disasters, which could stretch emergency management resources like Hurricane Andrew or the four hurricanes in 2004. Moreover, understanding environmental concerns has an effect on other areas of disaster response such as the evacuation procedures for citizens within those coastal areas (Wilson 1998). For instance, the Community Rating System (CRS), created by the National Flood Insurance Program, is used to understand which coastal community is the most vulnerable. Each one is given a score from 0 to 4,500 based off of 18 flood mitigation activities (Zahran et al. 2010). If someone lives within a high-scoring zone, then they are provided incentives on flood insurance to better protect their home. Examples of good practices for safe developments within communities is maintaining transportation infrastructure and elevating buildings above sea-levels or strengthening buildings against ground shaking or wind damage (Burby 1998) (Fig. 2.8).

Wilson (1998) performed research on coastal mandate compliance and found a couple factors for emergency managers to keep in mind, such as storm experiences, development patterns and planning capacity, with comprehensive emergency management plans. Florida learned its lesson with the unprecedented events of 1992

		Premium reduction		Florida localities	
Credit points	Class	Special Flood Hazard Area (SFHA) (%)	Non-SFHA (%)	Number of Florida localities	Percentage of Florida localities
4500+	1	45	10	0	0
4000–4499	2	40	10	0	0
3500–3999	3	35	10	0	0
3000–3499	4	30	10	0	0
2500–2999	5	25	10	6	2.9
2000–2499	6	20	10	13	6.2
1500–1999	7	15	5	60	28.6
1000–1499	8	10	5	97	46.2
500–999	9	5	5	34	16.2

Fig. 2.8 Overview of community rating system (Zahran et al. 2010)

and 2004. For Hurricane Andrew, the flooding was not the cause of $30 billion in losses- it was the wind gusts. By acknowledging the unique environmental factors, managers can create a more holistic plan to become more aware of how natural disasters will affect their communities (Fig. 2.9).

Florida embraces all hazard perspectives in disaster management. Within emergency management, increased attention has been placed on an all-hazards perspective to increase the resiliency and capability of each community. This perspective builds on the traditional model, which focuses on four phases to emergency management, and acknowledges the need for adaptation depending on unique needs for local, state and federal components (McEntire 2007). The four phases, consisting of mitigation, preparedness, response and recovery, align with the National Preparedness Goal of: "A secure and resilient nation with the capabilities required across the whole community to prevent, protect against, mitigate, respond to, and recover from the threats and hazards that pose the greatest risk" (Federal Emergency Management Agency 2014a). The following paragraphs provide a brief description of each phase.

2.2.8 Mitigation

The area of mitigation is deemed as one of the two most important priorities for emergency management personnel (EMP) as it focuses on prevention and reduction of potential impact (McEntire 2007). Kapucu and Garayev (2013) analyzed the area of mitigation and discovered three main activities. These activities incorporate EMPs targeting the threat to: (a) change the nature; (b) decrease vulnerability; and (c) reduce exposure. Overall, mitigation joins with preparedness to achieve goals, which are proactive in nature and benefit communities, by:

• "Identifying cost effective actions for risk reduction that are agreed upon by stakeholders and the public
• Focusing resources on the greatest risks and vulnerabilities

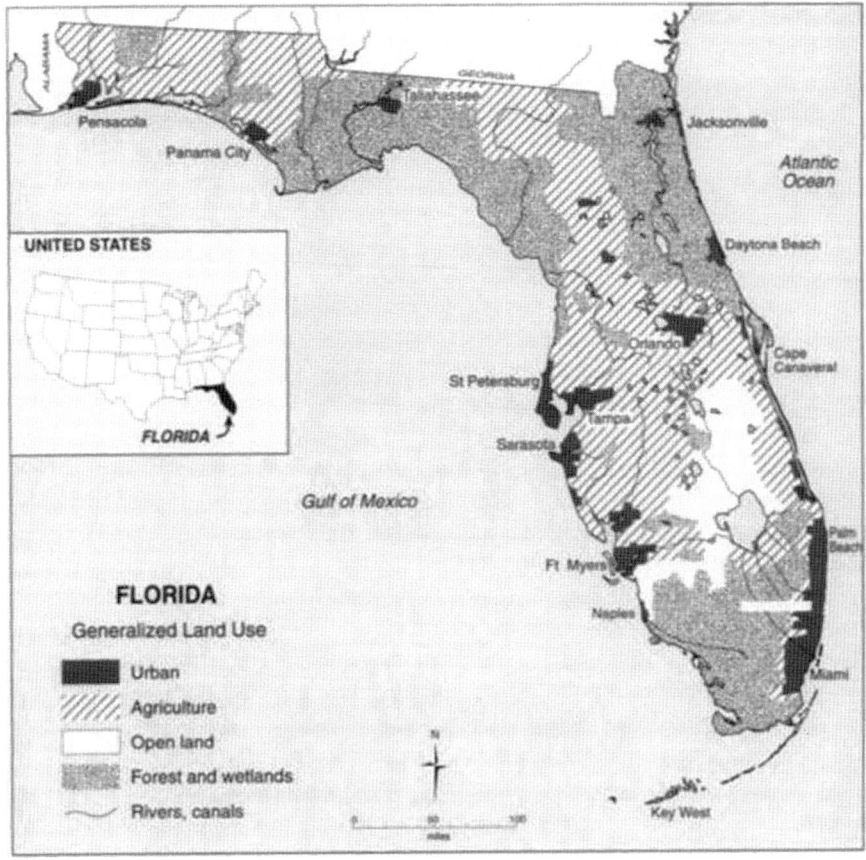

Fig. 2.9 Generalized land use planning (May et al. 1996)

- Building partnerships by involving people, organizations, and businesses
- Increasing education and awareness of hazards and risk
- Communicating priorities to state and federal officials
- Aligning risk reduction with other community objectives" (Federal Emergency Management Agency 2014b).

Florida uses several methods for disaster mitigation: State planning mandate requires counties (local government) to mitigate hazards, identify vulnerability using land use planning and development control mechanisms, and counties prepare comprehensive emergency management plans; additional state regulations for coastal region construction; and technical assistance and financial support for local mitigation initiatives. The Florida Division of Emergency Management and the Department of Environmental Protection are responsible in implementing mitigation strategies (Schapley and Schwartz 2014).

2.2.9 Preparedness

In regards to preparedness, this area is stated to be the second highest priority for emergency management personnel as it involves increasing the readiness for potential disasters and hazards (McEntire 2007). FEMA (2014b) generated a National Preparedness System to assist communities in the aforementioned goals along with the phase-specific components of planning, organizing/equipping, training, exercising, and evaluating/improving. Moreover, the system provides tools for risk assessment and creating operation plans and procedures.

2.2.10 Response

In the immediate aftermath of a disaster or hazard, emergency management personnel, along with affected local, state, and federal communities, engage in various response activities. These activities are important in safeguarding life and property (McEntire 2007). According to the United Nations (2008), *response* incorporates two objectives: "(1) increasing the capacity to predict, monitor, and reduce or avoid possible damage or addressing potential threats and (2) strengthening preparedness for response to a disaster or assist those who have been adversely affected" (p. 31). Moreover, it is important to recognize that response not only occurs immediately, but also incorporates short-term and long-term situations.

2.2.11 Recovery

Once the response efforts for a disaster or hazard are underway, it is time for emergency management personnel, along with affected community members, to resume operations and return to the predisaster, or improved, conditions (McEntire 2007). The National Disaster Recovery Framework assists affected communities achieve a well-managed recovery by promoting a flexible structure for decision-making and coordination, community engagement, financial management and resilience rebuilding (Federal Emergency Management Agency 2014c). In addition, this phase incorporates many short and long-term components and benefits through assessment, evaluation and redevelopment planning.

2.3 Emergency Management in Rural Communities

When analyzing the information regarding emergency management, there appears to be a connection between rural communities and areas of vulnerability. Several marginalization factors include limited economic, political, social and human resources (Morrow 1999). Economic disparities can be seen in the preparation phase where the citizenry cannot afford to purchase supplies in the event of a disaster. Furthermore, proactive planning can come to a halt if the community is unable to pay or perform services needed.

In the wake of a disaster, rural communities have the greatest chance of encountering issues in regards to response. Brennan and Flint (2007) found the citizens shouldering most of the responsibility for meeting the basic needs of affected individuals due to the existing disconnect between the community itself and local, state and federal officials. Moreover, there is a geographic issue as well due to the distance between rural and urban communities (Kapucu et al. 2013a, b). One avenue to battle this issue is through the creation of a Community Emergency Response Team (CERT) where emergency managers train local citizens to be the first line of defense. However, the question is whether the generation of a CERT enough?

With the unique make-up of each rural area, there are challenges to identifying and relying on specific members to be the main response and recovery team. In tandem with CERT, Brennan and Flint (2007) recommend bolstering emergency management policies and procedures with in-depth outlines of responsibilities on all levels of government. A positive attribute, to assist in the preparation and mitigation efforts of planning, is the existence of networks and alliances, which can be strengthened to heighten the community's capacity for disaster response (Kapucu et al. 2013a, b; Kleinberg 2014).

2.4 Conclusion

For a community to increase their capacity for resilience, they must prioritize emergency management and hazard planning policies while also creating supportive administrative structures (Hawkins and Knox 2014). Disaster emergency management response in Florida improved by trial and error after Hurricane Andrew and the 4 hurricanes that took place in 2004. As a result better policies were put in place, including better response frameworks that increased communication between organizations. Financial accountability was promoted and community coordination and collaboration took place to help eliminate the numbness generated by underestimation and under-preparedness (Kapucu 2008). An all hazard's approach was adopted which improved multi-sector collaboration and resulted in better comprehensive emergency management plans. In addition, an awareness of environmental concerns was promoted and rating systems were put in place for flooding insurance. Yet, with all these improvements rural communities still have response problems, particularly as a result of the physical distance between the community and emergency management personnel. To combat these issues the promoting of forming Community Emergency Response Teams (CERTS) has proven to increase the involvement of the community in disaster response.

Overall, while disasters bring about devastating social and economic consequences, they also have the potential to "shock" the community and promote better policies and disaster response frameworks, as the case of Florida illustrates. Nonetheless, communities do not need to wait to have a disaster situation to address these issues and promote disaster resilience. The lessons from Florida can served as a guiding light for communities wishing to be disaster resilience.

References

Adams, D. R. (1997). Local government financial emergencies Legislation in Florida: Definitions and reporting responsibilities. *Government Finance Review, 13*(3), 19–22.

Adams, D. R. (2014). Local government financial emergencies legislation in Florida: Definitions and reporting responsibilities. *Government Finance Review, 3*, 19.

Agranoff, R., & McGuire, M. (2003). *Collaborative public management: New strategies for local governments*. Washington, DC: Georgetown University Press.

American City and County. (2001). Disaster preparedness: Cities, county cooperate in hazard plan. *Issues & Trends*, 12–14.

Birkland, T. A. (2007). *Lessons of disaster: Policy change after catastrophic events*. Washington, DC: Georgetown University Press.

Brennan, M. A., & Flint, C. G. (2007). Uncovering the hidden dimensions of rural disaster mitigation: Capacity building through community emergency response teams. *Southern Rural Sociology, 22*(2), 111–126.

Brody, S. D., Zahran, S., Highfield, W. E., Bernhardt, S. P., & Vedlitz, A. (2009). Policy learning for flood mitigation: A longitudinal assessment of the community rating system in Florida. *Risk Analysis, 29*(6), 912–929.

Burby, R. J. (Ed.). (1998). *Cooperating with nature: Confronting natural hazards with land-use planning for sustainable communities*. Washington, DC: John Henry Press.

Chang, K. (2012). *Understanding cross-sector collaboration in emergency management: The dynamics of vertical and horizontal networks*. Electronic theses, treatises and dissertations. Paper 4765.

Choi, S. O. (2004). Emergency management growth in the state of Florida. *State & Local Government Review, 36*(3), 212–226.

Collier, M. C. (2012). Fostering resilience. *Trusteeship/Association of Governing Boards of Universities and Colleges, 20*(4), 41.

Deyle, R. E., Chapin, T. S., & Baker, E. J. (2008). The proof of the planning is in the platting: An evaluation of Florida's hurricane exposure mitigation planning mandate. *Journal of the American Planning Association, 74*(3), 349–370.

Dynes, R. (1970). *Organized behavior in disaster*. Lexington: Heath-Lexington Books.

Federal Emergency Management Agency (FEMA). (2011). *Government budgets: Long term trends and drivers and their implications for emergency management*. Washington, DC: FEMA.

Federal Emergency Management Agency. (2014a). *National preparedness goal*. Retrieved from http://www.fema.gov/national-preparedness-goal

Federal Emergency Management Agency. (2014b). *Multi-hazard mitigation planning*. Retrieved from http://www.fema.gov/multi-hazard-mitigation-planning

Federal Emergency Management Agency. (2014c). *National disaster recovery framework*. Retrieved from http://www.fema.gov/media-library/assets/documents/24647?fromSearch=fromsearch&id=5124

Hall, R. (2011). *Smart practices in building interorganizational collaborative capacity to strengthen the Florida comprehensive disaster management enterprise*. Monterey: Naval Postgraduate School.

Hawkins, C., & Knox, C. C. (2014). Disaster events and policy change in Florida. In N. Kapucu & K. T. Liou (Eds.), *Disaster & development: Examining global issues and cases* (pp. 111–128). New York: Springer.

Hughes, T. (2012). The evolution of federal emergency response since hurricane Andrew. *Fire Engineering, 165*, 90–4.

Kapucu, N. (2006). Interagency communication networks during emergencies: Boundary spanners in multi-agency coordination. *The American Review of Public Administration, 36*(2), 207–225.

Kapucu, N. (2008). Collaborative emergency management: Better community organizing, better public preparedness and response. *Disasters: The Journal of Disaster Studies, Policy, and Management, 32*(2), 239–262.

Kapucu, N. (2012a). Disaster resilience and adaptive capacity in Central Florida, US, and in Eastern Marmara region, Turkey. *Journal of Comparative Policy Analysis: Research & Practice, 14*(3), 202–216.

Kapucu, N. (2012b). Disaster and emergency management systems in urban areas. *Cities: The International Journal of Urban Policy and Planning, 29*(s1), 41–49.

Kapucu, N., & Garayev, V. (2013). Mitigation and emergency management. In A. Jerolleman & J. Kiefer (Eds.), *Natural hazard mitigation* (pp. 19–41). New York: CRC Press.

Kapucu, N., & Özerdem, A. (2013). *Managing emergencies and crises*. Boston: Jones & Bartlett Publishers.

Kapucu, N., & Van Wart, M. (2008). Making matters worse: Anatomy of leadership failures in catastrophic events. *Administration & Society, 40*(7), 711–740.

Kapucu, N., Berman, E., & Wang, X. (2008). Emergency information management and public disaster preparedness: Lessons from the 2004 Florida hurricane season. *International Journal of Mass Emergencies and Disasters, 26*(3), 169–197.

Kapucu, N., Augustin, M., & Garayev, V. (2009). Interstate partnerships in emergency management: Emergency management assistance compact (EMAC) in response to catastrophic disasters. *Public Administration Review, 69*(2), 297–313.

Kapucu, N., Hawkins, C. V., & Rivera, F. I. (2013a). Disaster preparedness and resilience for rural communities. *Risk, Hazards & Crisis in Public Policy, 4*(4), 215–233.

Kapucu, N., Hawkins, C., & Rivera, F. (Eds.). (2013b). *Disaster resiliency: Interdisciplinary perspectives*. New York: Routledge.

Kerjen, E. O. M. (2008). Toward a new risk architecture: The question of catastrophe risk calculus. *Social Research, 75*(3), 819–854.

Kleinberg, E. (2014). Emergency managers stress the need for social media volunteers. *McClatchy News*. Retrieved from http://www.emergencymgmt.com/disaster/Emergency-Managers-Stress-Need-Social-Media-Volunteers.html

Legislative Committee on Intergovernmental Relations. (2007). Review of the emergency management, preparedness, and assistance trust fund. *Interim Report*, Tallahassee.

Lindell, M. K., & Meier, M. J. (1994). Effectiveness of community planning for toxic chemical emergencies. *Journal of American Planning Association, 60*(2), 222–234.

MapWise. (2013, October 7). *Florida counties map*. Retrieved from http://www.mapwise.com/maps/florida/florida-counties-map.php

May, P. J., Burby, R. J., Ericksen, N. J., Handmer, J. W., Dixon, J. E., Michaels, S., & Smith, D. I. (1996). *Environmental management and governance: Intergovernmental approaches to hazards and sustainability*. New York: Routledge.

McEntire, D. A. (2007). *Disaster response and recovery: Strategies and tactics for resilience*. Hoboken, NJ: Wiley.

Mittler, E. (1997). *A case study of Florida's emergency management since Hurricane Andrew*. Boulder: Natural Hazards Research and Applications Information Center, Institute of Behavior Science, University of Colorado. Retrieved from http://www.colorado.edu/hazards/publications/wp/wp98.html

Morrow, B. H. (1999). Identifying and mapping community vulnerability. *Disasters, 23*(1), 1–18.

Mushkatel, A. H., & Weshcler, L. F. (1985). Emergency management and the intergovernmental system. *Public Administration Review, 45*(Special Issues), 49–56.

National Academy of Public Administration (NAPA). (1993). *Coping with catastrophe: Building an emergency management system to meet people's needs in natural and manmade disasters.* Washington, DC: NAPA.

Olson, R. S., & Gawronski, V. T. (2010). From disaster event to political crisis: A "5C+A" framework for analysis. *International Studies Perspectives, 11*, 205–221.

Rosenfeld, E. (2006). Dealing with disasters: Are we prepared? *Insurance Advocate, 117*, 19–26.

Rubin, C. B. (2013). *Emergency management: The American experience 1900–2005.* Fairfax: PERI.

Schapley, P. M., & Schwartz, L. (2014). Coastal hazard mitigation in Florida. In A. Farazmand (Ed.), *Handbook of crisis and emergency management* (pp. 771–789). Boca Raton: CRC Press.

Scott, R. (2014). *Gov. Scott signs legislation that creates sales tax holiday for hurricane supplies.* Retrieved from http://www.flgov.com/2014/05/12/gov-scott-signs-legislation-that-creates-sales-tax-holiday-for-hurricane-supplies/

Stallings, R. A., & Quarantelli, E. L. (1985). Emergent citizen groups and emergency management. *Public Administration Review, 45*(Special Issue), 93–100.

Stripling, M. (2013). *Managing chaos: The disaster planner's handbook in eight parts.* Retrieved from http://www.nyc.gov/html/doh/downloads/pdf/em/mc-disaster-handbook.pdf

Sylves, R. (2008). *Disaster policy and politics: Emergency management and homeland security.* Washington, DC: CQ Press.

Twigg, D. K. (2012). *The politics of disaster: Tracking the impact of Hurricane Andrew.* Orlando: University Press of Florida.

United Nations (UN). (2008). *Disaster preparedness for effective response: Guidance and indicator package for implementing priority five of the Hyogo framework.* Retrieved from http://www.un.org.np/attachments/disaster-preparedness-effective-response-guidance-and-indicator-package-implementing-pri

Wamsley, G. L., & Shroeder, A. D. (1996). Escalating in a quagmire: The changing dynamics of emergency management policy subsystem. *Public Administration Review, 56*(3), 235–244.

Waugh, W. L., Jr. (1994). Regionalizing emergency management: Counties as state and local government. *Public Administration Review, 54*(3), 253–258.

Waugh, W. L., Jr. (Ed.). (2006). Shelter from the storm: Repairing the national emergency management system after Hurricane Katrina. Special issue of *The Annals of the American Academy of Political and Social Science, 604*, 256–272.

Wilson, H. W. (1998). Local government compliance with state planning mandates: The effects of state implementation in Florida. *Journal of the American Planning Association, 64*(4), 457–469.

Wilson, J., & Oyola-Yemaiel, A. (2001). The evolution of emergency management and the advancement towards a profession in the United States and Florida. *Safety Science, 39*, 117–131.

Zahran, S., Brody, S. D., Highfield, W. E., & Vedlitz, A. (2010). Non-linear incentives, plan design, and flood mitigation: The case of the federal emergency management agency's community rating system. *Journal of Environmental Planning and Management, 53*(2), 219–239.

Chapter 3
Geography and Resilience

Abstract As with most response and recovery efforts, context is everything. Differences in social, political, geographic and economic structures necessitate attention. For Florida, the geographic composition brings about various disaster and hazard vulnerabilities. Moreover, the field of emergency management becomes impacted in resilience capabilities. The belief of a one-size-fits-all approach decreases effectiveness of mitigation efforts and negatively impacts each community (Cutter SL, Emrich CT, ANN Am Acad Polit Soc Sci 604:102–112, 2006; Cutter SL, Barnes L, Berry M, Burton C, Evans E, Tate E, Webb J, Glob Environ Chang 18(4):598–606, 2008; Schwab J, Hazard mitigation: integrating best practices into planning. Retrieved from http://www.fema.gov/media-library/assets/documents/19261, 2011). This chapter looks to understand the impact of geography on disasters and hazards while also increasing awareness of Florida-specific vulnerability and resiliency issues.

Keywords Geography • Disaster resilience • Hazards • Vulnerability • Florida

3.1 Geography, Disasters, and Hazards

Examining the research of disasters and hazards leads to the analysis of cause and effects. Geographers evaluated natural hazards early on to discover patterns and spatial distributions of physical processes along with the impact of specific events (Montz and Tobin 2011). In 1945, a significant change occurred with the publication of Gilbert White's research of natural hazards in relation to solving societal issues. This landmark not only affected geographers in general, but it also revealed the "truly inter-disciplinary concerns involving virtually all the social and natural sciences, health interests and professional programs" (Montz and Tobin 2011, p. 1).

In areas like Florida, full of geographic nuances like the coastal boundary and climate differentials, preparation, mitigation, response and recovery efforts become greatly affected (Tobin 1999; Burby et al. 2000; Montz and Tobin 2011; Schapley and Schwartz 2014; Schwab 2011). Natural hazards have been estimated to cost US citizens approximately $500,000,000 per week. Coastal reconstruction, specifically, has resulted in billions of tax dollars being used. "Under the law, the federal

© Springer International Publishing Switzerland 2015

F.I. Rivera, N. Kapucu, *Disaster Vulnerability, Hazards and Resilience*,
Environmental Hazards, DOI 10.1007/978-3-319-16453-3_3

government committed more than $80 billion to disaster recovery from 2004 to 2011" (Gillis and Barringer 2012, para. 13). In an effort to reduce predicted and residual costs of these events, programs like land-use planning are promoted to help a community understand the strengths and weaknesses in the landscape. These plans not only help articulate goals, principles, and strategies, but they support sustainability and adaptive capacity. However, administrators must keep in mind the potential effects of climate change as long-term, land-use plans are based on past events and predicted future situations (Tobin 1999; Daniels and Daniels 2003; Beatley 2009; Frazier et al. 2010a, b; Dannenberg et al. 2011). Mitigation, deemed a cornerstone of emergency management, is deeply connected to the planning efforts due to its focus on alleviating long-term impacts of disasters and hazards to build resiliency and sustainability (Tobin 1999; Schwab 2011).

To begin the planning process, stakeholders need to be identified before administrators use assessment measures to generate a factual foundation for decision-making (Frazier et al. 2010a, b). Some potential categorization and identification of stakeholders incorporates:

- Business, including representatives from local chambers of commerce, insurance companies, restaurants, tourist accommodations, and retail trade;
- Environment, including estuary program coordinators and marine researchers;
- Emergency Management and Infrastructure, including public works managers, county health officials, recovery planners, and county emergency managers;
- Government Officials, including city managers, vice mayors, city commissioners, and sustainability officers; and
- Planning, including city, county, and regional planning officials (Frazier et al. 2010b, pp. 509–510).

Three levels, incorporated within the analysis, consist of hazard identification, vulnerability assessment, and risk analysis (Burby et al. 2000; Beatley 2009). A first step in the overall assessment process can include the Spatial Hazard Event and Loss Database for the United States (SHELDUS) (NAS 2012). This database is as close to a full compilation of disaster events starting in the 1960s. Once this information has been compiled, the next step can incorporate:

> Vulnerability assessment combines the information from hazard identification with an inventory of the existing (or planned) property and population at risk. It provides information on who and what are vulnerable to a natural hazard within the geographic areas defined by hazard identification and can estimate damages and casualties that will result from various intensities of the hazard. Risk analysis includes a full probability assessment of the various intensities of a hazard as well as probability assessment of impacts on structures and populations. (Burby et al. 2000, p. 102)

Once the assessment is complete, the results can be translated into support for needed changes. For example, if the data shows that a county has a higher risk of structural damage from hurricane force winds, then policy makers can use the results to encourage building regulations for future development and preventative strategies for current buildings. Through these proactive approaches, communities are able to bring about awareness, produce change, and offset consequences (Burby et al. 2000; Frazier et al. 2010a, b; Schwab 2011; Dannenberg et al. 2011).

3.2 Geography and Disaster Vulnerability

Once a specific geographic area has been examined in regards to potential for disasters and hazards then administrators can narrow down the information to focus on vulnerability of the communities. Vulnerability is the culmination of results from resource availability/distribution along with public policies (Chakraborty et al. 2005; Montz and Tobin 2011; National Oceanic and Atmospheric Administration & the Environmental Protection Agency 2011). This human-induced situation is attributed to increasing or decreasing a hazard or disaster's effect.

Speaking to Florida's unique geographic climate, Cutter and Emrich (2006) broadly connected social vulnerabilities to geographic landscapes like coastal regions. Social vulnerabilities are the products of inequalities and impact various social groups who are deemed more susceptible to hazards. As the Florida's population grows, the social variations increase as well (Tobin 1999). "Coastal counties now have significant year-round residents- many of them elderly retirees or service industry workers who keep the tourist industry afloat" (Cutter and Emrich 2006, p. 103). These variations affect gaps in economic disparities, which then affects the community's ability to prepare and recover when disasters strike.

With social vulnerability being connected to variables of socioeconomic status, gender, race and ethnicity, age, commercial and industrial development, employment loss, rural/urban, residential property, infrastructure and lifelines, renters, occupation, family structure, education, population growth, medical services, social dependence and special needs populations, Cutter et al. (2003) identified three overarching tenants within the research consisting of the built environment, the natural landscape and the social factor. By understanding how these tenants intermingle within a specific geographic area, administrators can better understand how their citizenry will be impacted by disasters and hazards through improvement in assessment efforts and justification of "the selective targeting of communities for mitigation based on good social science, not just political whim" (Cutter et al. 2003, p. 258).

To assist in research, the Social Vulnerability Index (SoVI) was created. This tool uses a multitude of variables signifying socioeconomic, demographic, and built environment aspects to determine level of resilience (Chakraborty et al. 2005; Cutter et al. 2008; Cutter and Finch 2008). Cutter et al. (2008) indexed the United States to determine SoVI of every county in the nation. A visual representation of this is seen in Fig. 3.1.

A specific case study example of social vulnerability for Florida is seen in Chakraborty et al. (2005) utilization of SoVI and geographical information systems (GIS) to determine evacuation needs for Hillsborough County. GIS is an important spatial analysis tool to evaluate parameters and unveil new or unsuspected relationships (Finkl 2000). Researchers selected this county due to its susceptibility to hurricanes, lightning, droughts, floods, and more. After analysis, researchers determined significant differences affecting emergency management efforts. For instance, a number of coastal locations were deemed very susceptible to hazards, yet social vulnerability was relatively low. Conversely, there were areas of low geographic risk, but high social vulnerability.

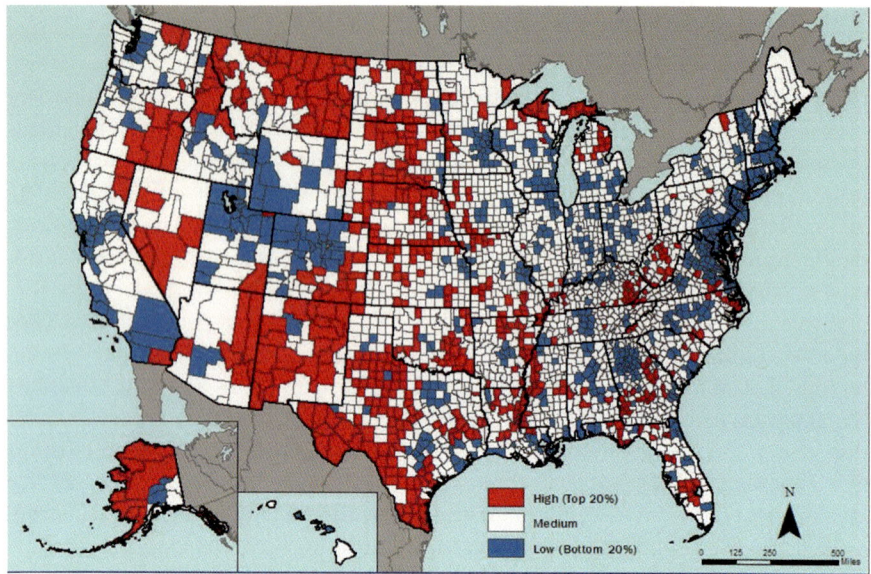

Fig. 3.1 Social Vulnerability Index (SoVI) of United States' counties (NAS 2012)

Placing geophysical risk and social vulnerability as connected entities only increases mitigation ability of a community, especially for a coastal community (Finkl 2000; Boruff et al. 2005; Chakraborty et al. 2005; Schapley and Schwartz 2014; Schwab 2011). "These results, then, call for a two-pronged approach to evacuation planning, one prong concentrating on high-risk areas and the other on particular needs of populations in particular areas, regardless of the magnitude of geophysical risk" (Chakraborty et al. 2005, p. 32). In conjunction with the intersection, it is imperative for emergency management personnel to continuously re-evaluate each area, as these measures are not static (Finkl 2000). There will always be demographic transitions and climate changes can only be predicted to an extent (Chakraborty et al. 2005; Cutter and Finch 2008).

3.3   Geography and Disaster Resilience

As the field of emergency management continues to develop, so do preparedness methods and strategies. Among federal agencies, a noticeable shift occurred when the rhetoric relating to disasters and hazards went from vulnerability to resilience. Cutter et al. (2008) believe the transition encompasses a more positive and proactive perspective to communities. Within the definitions themselves, vulnerability lends itself to potential for harm and sensitivies of a system verses resilience, which speaks to adaptive processes to respond and recover.

The Committee on Increasing National Resilience to Hazards and Disasters, the Committee on Science, Engineering, and Public Policy, and the National Academies (hereafter referred to as NAS) (2012) prepared a new paradigm of planning, a vision that was generated by incorporating several components. The first was to simply take responsibility for disaster risk followed by assessing past losses, due to disasters, to support long-term planning efforts. Also, communities are responsible for utilizing tools or metrics to monitor progress and build capacity. Lastly, administrators need to understand the context of current policies and practices along with identifying and clarifying roles and responsibilities (NAS 2012).

In reviewing standard tools and measures to analyze resilience, researchers discovered an issue as most models involve frameworks to capture infrastructure, but do not account for antecedent social factors (Boruff et al. 2005; Cutter et al. 2008). Therefore, Cutter et al. (2008) developed a theoretically driven disaster resilience of place (DROP) model to present the connection between resilience and vulnerability in terms of a cumulative effect. The total impact of a hazard or disaster is seen to encompass event characteristics, antecedent conditions, and coping responses (see Fig. 3.2). Within the indicators, geography heavily influences the ecological dimension as resilience is affected by spatiality and biodiversity.

Shortly after the DROP model, a composite indictor based on five sub-indexes was created and titled the Baseline Resilience Indicator for Communities (BRIC). Compilation of economic, institutional, social, community resilience, and infrastructure were included, but ecological, or geographic, nuances were not originally. However, to see the indicators in action, administrators can utilize HAZUS, which is a modeling tool produced by the Federal Emergency Management Agency (FEMA 2003; NAS 2012). This program runs scenarios or actual events to see how a community is impacted. The data further supports long-term planning and increases capacity for resiliency. If interested in other national models and frameworks for resilience measurements, researchers can use the: Community Assessment of Resilience Tool; Community Resilience System; the Toolkit for Health and Resilience in

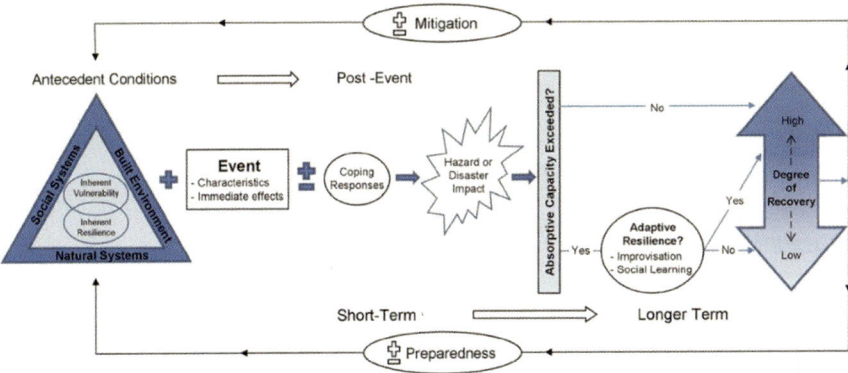

Fig. 3.2 Schematic representation of the Driven Disaster Resilience of Place (DROP) model (Cutter et al. 2008)

Vulnerable Environments; the Community Resilience Model; the Resilience Capacity Index, the Community Disaster Resilience Index; and the Center for Risk and Economic Analysis of Terrorism Events Economic Resilience Index (NAS 2012).

Regardless of process, it is helpful to keep the nation's resiliency goals in mind. A key element is to understand risks to the community and create policies to help protect critical functions. The risks mentioned within this element include flood plains and coastal wetlands, which are aspect Florida grapples with during disasters and hazards.

3.4 Coastal Disaster Resilience

As the focus continues to increase on coastal resilience, it is natural to see a similar development in arenas outside of policies. For example, the combination of GIS and remote sensing generates specialized coastal maps to support rationale of resilience policies by highlighting regions deemed more vulnerable to disasters and hazards along with urban development and sprawl (Finkl 2000; Boruff et al. 2005; Frazier et al. 2010a, b; Florida Department of Economic Opportunity 2014). These maps are especially useful in land-use planning as natural hazards, such as sea level rise, increasingly impact coastal communities.

In regards to specific guides, the Coastal-Marine Ecosystem-Based Management Tools Network generated a guide specifically for coastal management administrators. Utilizing adaptation planning, the network synthesizes the planning process into six steps: scoping the problem, inventorying and collecting data, analyzing the information, developing a plan, developing strategies, and then monitoring and evaluating outcomes (NatureServe 2013). Staying focused on coastal regions, tools promoted within the guide consist of web-based portals and software deemed functional, applicable to diverse range of geographies, and user-friendly. For a brief view of the tools in regards to sector specialization, refer to Table 3.1.

The aforementioned tools are especially useful when analyzing hazards such as sea-level rise, which is a growing concern for coastal communities due to its

Table 3.1 Tools by sector (NatureServe 2013)

	Natural resources	Agriculture	Built environment	Transportation	Energy
CanVis	X		X		
SLR Viewer	X		X	X	
SoVI			X		
HAZUS			X	X	X
OpenNSPECT	X	X	X		
SLAMM	X				
InVEST	X	X			X
CCVI	X				
CommunityViz	X	X	X	X	X
NatureServe Vista	X	X	X	X	X

connection with climate change (Murley et al. 2008; Frazier et al. 2010a, b). To date, the impact of this situation is seen in inland migration of residents and changes in inundation zones due to physical landscape changes (Frazier et al. 2010a, b). Moreover, the Intergovernmental Panel on Climate Change has predicted an increase of 9–23 in. of sea-level rise by the next century. Though the numbers may seem small, their impact will only exacerbate potential for increased hurricane intensities along with added impact in severe droughts, torrential flooding, and urban development concerns (Murley et al. 2008). For a visual depiction, the SLR Viewer can be initialized as it stands for Sea Level Rise and Coastal Flooding Impact Viewer and allows researchers the ability to see potential impacts through mapping. Moreover, SLAMM, or the Sea Level Affecting Marshes Model, holds a similar viewing capability as the SLR Viewer, but is focused more on wetland conversation and shoreline modifications (NatureServe 2013).

Frazier et al. (2010a, b) investigated sea-level rise and storm surge on social vulnerabilities for Sarasota County, Florida. Through prediction of sea level rise due to climate changes, the researchers established significant increases to social vulnerability when examining geographic changes (see Fig. 3.3).

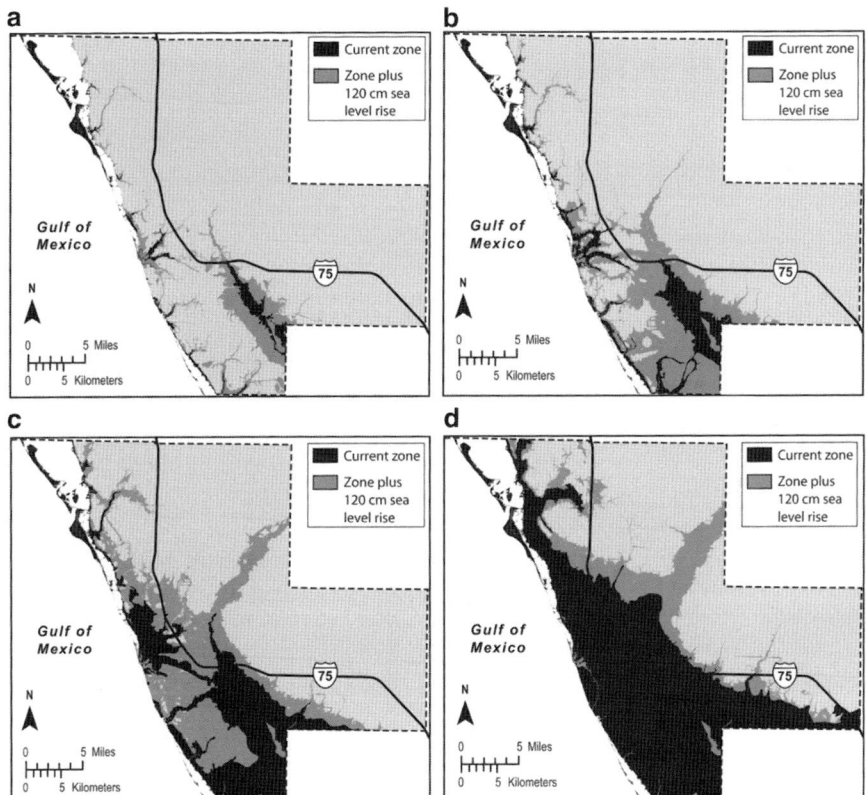

Fig. 3.3 Storm surge and sea level rise projections of Sarasota County. (**a**) Category 1 storm-surge hazard zones. (**b**) Category 2 storm-surge hazard zones. (**c**) Category 3 storm-surge hazard zones (**d**) Category 4 and 5 storm-surge hazard zones (Frazier et al. 2010b)

Much like the aforementioned DROP model, comprehensive planning can be undertaken by each community to articulate strategies for redevelopment (Florida Department of Community Affairs and Florida Department of Emergency Management 2010; NAS 2012). "The combined factors of hurricane storm-surge inundation, the potential amplifying effect of sea level rise on inundation zones, and the continuing development of the coast indicate a pressing need for coastal communities to conduct comprehensive vulnerability assessments as they develop long-term land-use plans" (Frazier et al. 2010a, b, p. 2). A post-disaster redevelopment plan is mandatory for each coastal region of Florida and is also an optional process for inland regions. It is critical for Florida's counties and municipalities to be knowledgeable of generated and/or adapted statutes and rules.

Moreover, it is important to be cognizant of how to articulate the technical information for community members so they understand cause, effect and impact (Montz and Tobin 2011; Ruppert 2011; Schwab 2011; NAS 2012). Regarding the written planning document, some suggested elements consist of standards and codes, performance metrics and rating systems, community organizations, local capacity, education and communication, and resource management (NAS 2012).

3.5 Other Hazards

In addition to hazards in coastal regions the geographic composition of Florida brings about other natural hazards including heat waves, floods, and storm surges. Heat waves are particularly relevant to Florida's elderly population as they tend to worsen pre-existing health conditions and are linked to elevated rates of mortality (Comrie 2007). Previous studies have found elevated heat mortality risk in the Florida cities of Jacksonville and Miami in comparison to other cities in the U.S. east (Curriero et al. 2002). Risk models, such as the Extreme Value Analysis Point Process approach (Keellings and Waylen 2014), provide useful tools to identify heat waves areas that can be used to ameliorate its effects in the face of continuous climate changes.

Another geographical hazard for Florida is flooding, mainly due to its low elevation, large coastal population, and frequent storm events. In addition "the combination of rapid population growth and related development, the alteration of hydrological systems through building and channelization, and large amounts of annual precipitation associated with a tropical and subtropical climate have made many local jurisdictions across the state vulnerable to repetitive flooding and flood damage" (Brody et al. 2007, p. 331). Indeed, Brody et al. 2007 in their study of floods in Florida, point to the importance of planning decision in curtailing the potential damage brought about by flooding events. As the authors eloquently concluded: "Precisely where we choose to develop and how we protect communities from natural hazards influences how much property damage floods produce. Carefully weighing the costs and benefits of these decisions thus becomes critical to building sustainable, resilient communities for future generations (Brody et al. 2007, p. 343)".

Related to flooding are the potential effects of storm surges in areas considered to be a risk from sea-level rise such as Miami and Palm Beach. Risk assessment models, such as the SLOSH model, indicate that the potential economic losses for storm surges will be devastating in the absence of preventive protections, such as building embankments or walls, scale down coastal development and beach erosion reduction plans (Genovese and Green 2014; Natural Weather Service National Hurricane Center 2014). SLOSH stands for Sea, Lake, and Overland Surge from Hurricanes. It is a computerized model developed by the National Weather Service (NWS) to estimate storm surge heights and winds resulting from historical, hypothetical, or predicted hurricanes. SLOSH is used by the National Hurricane Center (NHC) for the exclusive benefit of NWS, US Army Corps of Engineers (USACE), and Emergency Management personnel. SLOSH is the primary model used by the Federal Emergency Management Agency (FEMA), the National Oceanographic and Atmospheric Administration (NOAA), and USACE. It is also the basis for Hurricane Evacuation Studies (HES) (http://slosh.nws.noaa.gov/sloshPub/SLOSH-Display-Training.pdf).

All these hazards and assessments models have provided significant information that has impacted the discussion and implication of resilience policies which we discuss in the next section.

3.6 Coastal Disaster Resilience Policies and Implementation in Florida

When discussing the roles and responsibilities of federal and state government officials, assistance can only extend so far. These governing entities are there to assist in local efforts, yet they do not have the capacity to research each community and create appropriate plans and policies. Therefore, it is up to local governments to take ownership. "Resilience to disasters rests on the premise that these multiple systems are robust, and that the system components work in concert and in such a way that the interdependencies provide strength during a disaster event" (NAS 2012, p. 167).

Florida is one state that has empowered their administrators in planning efforts. The Local Government Comprehensive Planning and Land Development Regulation Act is not only responsible for requiring all counties and municipalities to adopt a comprehensive plan, but it was influential in requiring every coastal community to create a post-disaster redevelopment plan (PDRP) (FDCA and FDEM 2010). To establish a PDRP, responsible officials can utilize the process shown in Fig. 3.4.

Once the plan has been drafted, it is the responsibility of each committee member to evaluate specific needs for their community. Nuances can be discovered by analyzing the following elements: Future land use; Coastal management; Conservation; Local mitigation strategy (hazard analysis, vulnerability and risk assessment, and goals and objectives); Long-range transportation plan (multimodal, needs plan, and cost-feasible plan); and other local/regional plans (transfer of development rights program, land acquisition program, land development codes, community visioning plans, and area-specific redevelopment plans) (FDCA and FDEM 2010) (Fig. 3.5).

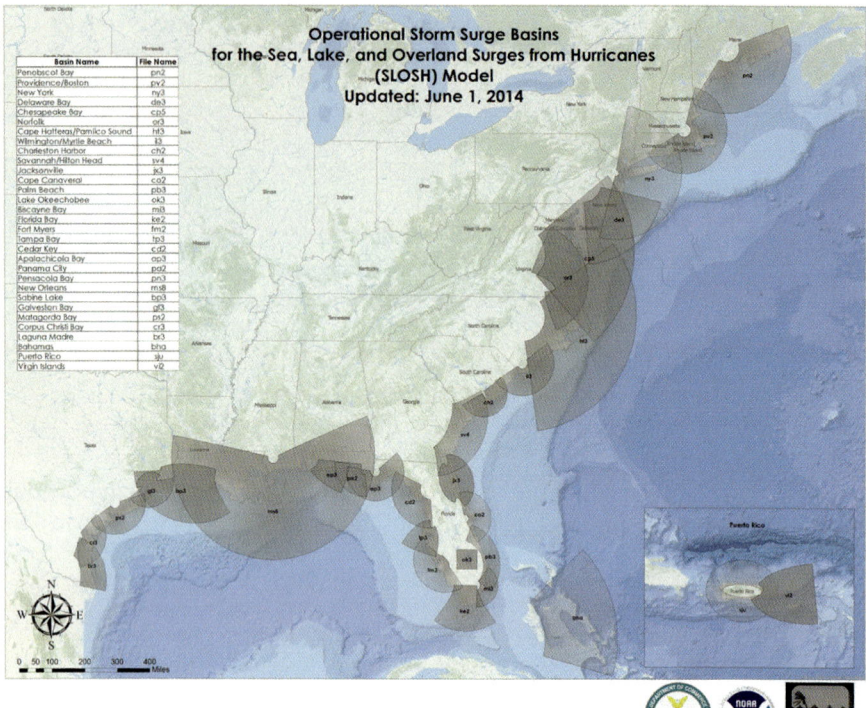

Basin Name	File Name
Penobscot Bay	pn2
Providence/Boston	pv2
New York	ny3
Delaware Bay	de3
Chesapeake Bay	cp5
Norfolk	or3
Cape Hatteras/Pamlico Sound	ht3
Wilmington/Myrtle Beach	il3
Charleston Harbor	ch2
Savannah/Hilton Head	sv4
Jacksonville	jx3
Cape Canaveral	cc2
Palm Beach	pb3
Lake Okeechobee	ok3
Biscayne Bay	mi8
Florida Bay	ke2
Fort Myers	fm2
Tampa Bay	tp3
Cedar Key	cd2
Apalachicola Bay	ap3
Panama City	pa2
Pensacola Bay	pn3
New Orleans	ms8
Sabine Lake	bp3
Galveston Bay	gl3
Matagorda Bay	pn2
Corpus Christi Bay	cr3
Laguna Madre	br3
Bahamas	bha
Puerto Rico	sju
Virgin Islands	vi2

Operational Storm Surge Basins for the Sea, Lake, and Overland Surges from Hurricanes (SLOSH) Model — Updated: June 1, 2014

Fig. 3.4   Operational Storm Surge Basins (http://www.nhc.noaa.gov/surge/slosh.php#SDISPLAY)

Focusing on the coastal management element, Rule 9 J-5.012 (3)(c)(5) of the Florida Administrative Code details the need for policies to "accomplish the following:

- Distinguish between immediate repair and cleanup actions needed to protect public health and safety and long-term repair and redevelopment activities;
- Address the removal, relocation, or structural modification of damaged infrastructure as determined appropriate by the local government but consistent with Federal funding provisions and unsafe structures;
- Limit redevelopment in areas of repeated damage; and
- Incorporate the recommendations of interagency hazard mitigation reports, as deemed appropriate by the local government, into the local government's comprehensive plan when it is revised during the evaluation and appraisal process" (FDCA and FDEM 2010, p. 3).

In conjunction with the redevelopment plan is Section 163.3178(2) of the Florida statutes dictating all coastal management elements having to be based on data, surveys, and studies. In addition, these resource plans are to eliminate unsafe

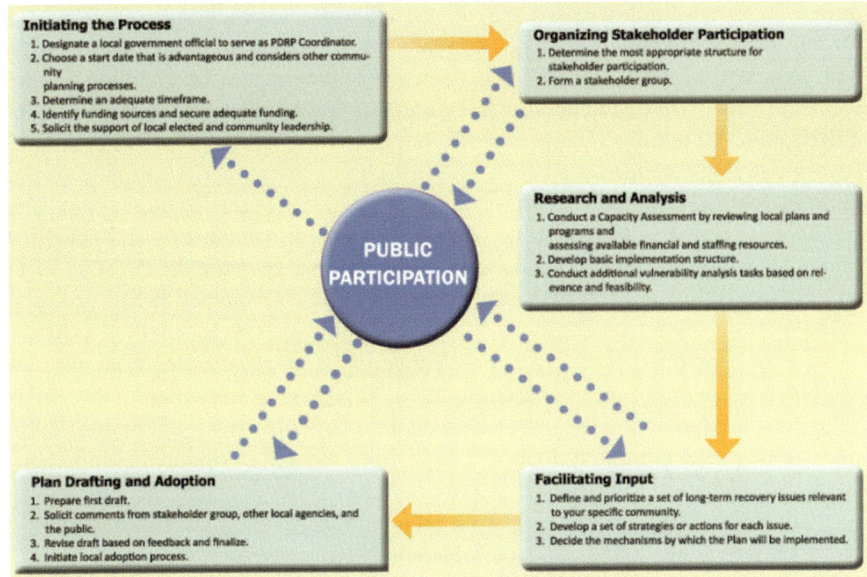

Initiating the Process
1. Designate a local government official to serve as PDRP Coordinator.
2. Choose a start date that is advantageous and considers other community planning processes.
3. Determine an adequate timeframe.
4. Identify funding sources and secure adequate funding.
5. Solicit the support of local elected and community leadership.

Organizing Stakeholder Participation
1. Determine the most appropriate structure for stakeholder participation.
2. Form a stakeholder group.

Research and Analysis
1. Conduct a Capacity Assessment by reviewing local plans and programs and assessing available financial and staffing resources.
2. Develop basic implementation structure.
3. Conduct additional vulnerability analysis tasks based on relevance and feasibility.

PUBLIC PARTICIPATION

Plan Drafting and Adoption
1. Prepare first draft.
2. Solicit comments from stakeholder group, other local agencies, and the public.
3. Revise draft based on feedback and finalize.
4. Initiate local adoption process.

Facilitating Input
1. Define and prioritize a set of long-term recovery issues relevant to your specific community.
2. Develop a set of strategies or actions for each issue.
3. Decide the mechanisms by which the Plan will be implemented.

Fig. 3.5 Post-disaster redevelopment plan process (FDCA and FDEM 2010)

development and adhere to Rule 9 J-5.012(2)(e) of the Florida Administrative Code, which states the necessary analysis of:

> Existing and proposed land use in coastal high-hazard areas; structures with a history of repeated damage in coastal storms; coastal or shore protection structures; infrastructure in coastal high-hazard areas; and beach and dune conditions. Measures which could be used to reduce exposure to hazards shall be analyzed, including relocation, structural modification, and public acquisition. Coastal high-hazard areas shall be identified and the infrastructure within the coastal high-hazard area shall be inventoried. The potential for relocating threatened infrastructure shall be analyzed. (FDCA and FDEM 2010, p. 138)

Preliminary attempts to guide Florida communities in creating policies incorporating climate change initiatives include Murley et al. (2008) framework promoting disciplined, comprehensive, purposeful, strategic and efficient adaptation policies. The framework consists of critical questions and policy options for comprehensive land-use planning, transportation and other infrastructure, coastal management, water resource management, and emergency preparedness. In addition, the researchers challenge administrators to "get the science right" as quantity and quality of the information "on climate change impacts and the effectiveness of coastal climate adaptation strategies, now and for many decades, means that we must make a fundamental commitment to fostering the highest quality innovation, information, and analysis in climate adaptation science" (Murley et al. 2008, p. 21).

To increase proactive approaches in Florida's coastal communities, the legislature passed a Community Planning Act (CPA) in 2011. This act incorporates adaptation planning for coastal hazards to increase awareness of sea level rise and improve resiliency through comprehensive planning. Supporting this endeavor is the National Oceanic and Atmospheric Administration (NOAA) who approved the Florida Coastal Management Program's (FCMP) initiative to examine statewide planning frameworks and establish best practices for integrating sea level rise efforts. The Department of Economic Opportunity (DEO) is conducting the 5-year initiative and has projected a two-phase implementation process. The first phase incorporates data collection of sea level rise research along with technical assistance resource identification. The second phase consists of model generation for sea level rise inundation and guideline creation. The hope is for Florida to holistically plan and increase their resiliency to sea level rise (Ruppert 2011; Department of Economic Opportunity 2014).

To determine effectiveness of the initiative, pilot studies were conducted. At the forefront of adaptation planning is Fort Lauderdale, only one of the communities who opted to be a pilot for the DEO initiative. Through this designation, the city is in the process of adopting the optional designation by integrating adaptation action areas (AAA) into their comprehensive plan. An AAA is "an optional comprehensive plan designation for areas that experience coastal flooding and that are vulnerable to the related impacts of rising sea levels for the purpose of prioritizing funding for infrastructure needs and adaptation planning" (City of Fort Lauderdale 2014).

The role taken by Fort Lauderdale is indicative of the collaborative nature of Florida counties in general. By offering to wade through the waters first, the state as a whole is able to adjust recommendations and alleviate any implementation stress for the other counties. Moreover, the more tangible connection shows the importance of acknowledging climate change and planning for its impact (Ruppert 2011). Even more, the cities and counties who initiate coastal management within their land use plans are able to provide samples to other regions to help in their development. For example, the City of Satellite Beach (2010) incorporated sea-level rise into their mitigation efforts and were able to predict the amount of loss to their environment along with data to help prioritize response initiatives alongside other planning goals.

3.6.1 Conclusion

Florida's geographic composition significantly impacts resiliency efforts due to vulnerability to various disasters and hazards. The field of emergency management needs the support of local government to understand these impacts and build community capacity. In this chapter we reviewed some of the tools used to identify hazards, assess vulnerabilities and analyze risks. These include, among others, the Spatial Hazard Event and Loss Database for the United States (SHELDUS), the Driven Disaster Resilience of Place (DROP), and the Social Vulnerability Index (SoVI). We also discussed some of the planning efforts taking place in Florida, including the Local Government Comprehensive Planning and Land Development Regulation Act, among others. Overall, the hope is to increase awareness of geographic nuances, especially coastal management, through identification, assessment and planning processes.

References

Beatley, T. (2009). *Planning for coastal resilience: Best practices for calamitous times.* Washington, DC: Island Press.

Boruff, B. J., Emrich, C., & Cutter, S. L. (2005). Erosion hazard vulnerability of US coastal counties. *Journal of Coastal Research, 21*(5), 932–942.

Brody, S. D., Zahran, S., Maghelal, P., Grover, H., & Highfield, W. (2007). The rising costs of floods: Examining the impact of planning and development decisions on property damage in Florida. *Journal of the American Planning Association, 73*, 330–345.

Burby, R. J., Deyle, R. E., Godschalk, D. R., & Olshansky, R. B. (2000). Creating hazard resilient communities through land-use planning. *Natural Hazards Review, 1*(2), 99–106.

Chakraborty, J., Tobin, G. A., & Montz, B. E. (2005). Population evacuation: Assessing spatial variability in geophysical risk and social vulnerability to natural hazards. *Natural Hazards Review, 6*(1), 23–33.

City of Fort Lauderdale. (2014). *Innovative pilot projects.* Retrieved from http://www.fortlauderdale.gov/gyr/climate_resiliency/pilot_projects.html

City of Satellite Beach. (2010). *Municipal adaptation to sea-level rise: City of Satellite Beach, Florida.* Retrieved from http://www.satellitebeachfl.org/Documents/Sea%20Level%20Rise%20-%20CRE%20Report%2007-18-10.pdf

Committee on Increasing National Resilience to Hazards and Disasters, Committee on Science, Engineering, and Public Policy, & The National Academies. (2012). *Disaster resilience: A national imperative.* Washington, DC: National Academies Press.

Comrie, A. (2007). Climate change and human health. *Geography Compass, 1*(3), 325–339.

Curriero, F. C., Heiner, K. S., Samet, J. M., Zeger, S. L., Strug, L., & Patz, J. A. (2002). Temperature and mortality in 11 cities of the eastern United States. *American Journal of Epidemiology, 155*, 80–87.

Cutter, S. L., & Emrich, C. T. (2006). Moral hazard, social catastrophe: The changing face of vulnerability along the hurricane coasts. *The ANNALS of the American Academy of Political and Social Science, 604*, 102–112.

Cutter, S. L., & Finch, C. (2008). Temporal and spatial changes in social vulnerability to natural hazards. *Proceedings of the National Academy of Sciences, 105*(7), 2301–2306.

Cutter, S. L., Boruff, B. J., & Shirley, W. L. (2003). Social vulnerability to environmental hazards. *Social Science Quarterly, 84*(2), 242–261.

Cutter, S. L., Barnes, L., Berry, M., Burton, C., Evans, E., Tate, E., & Webb, J. (2008). A place-based model for understanding community resilience to natural disasters. *Global Environmental Change, 18*(4), 598–606.

Daniels, T., & Daniels, K. (2003). *The environmental planning handbook.* Chicago: Planners Press.

Dannenberg, A., Frumkin, H., & Jackson, R. (Eds.). (2011). *Making healthy places: Designing and building for healthy, well-being, and sustainability.* Washington, DC: Island Press.

Department of Economic Opportunity (DEO). (2014). *Florida adaptation planning.* Retrieved from http://www.floridajobs.org/community-planning-and-development/programs/technical-assistance/community-resiliency/adaptation-planning

Federal Emergency Management Agency (FEMA). (2003). *SLOSH display training.* Retrieved from http://slosh.nws.noaa.gov/sloshPub/SLOSH-Display-Training.pdf

Finkl, C. W. (2000). Identification of unseen flood hazard impacts in southeast Florida through integration of remote sensing and geographic information system techniques. *Environmental Geosciences, 7*(3), 119–136.

Florida Department of Community Affairs & Florida Department of Emergency Management. (2010). *Post-disaster redevelopment plan.* Retrieved from http://www.floridadisaster.org/recovery/documents/Post%20Disaster%20Redevelopment%20Planning%20Guidebook%20Lo.pdf

Florida Department of Economic Opportunity. (2014). *Adaptation planning in Florida*. Retrieved from http://www.floridajobs.org/fdcp/dcp/AdaptationPlanning/AdaptationPlanninginFlorida.pdf

Frazier, T. G., Wood, N., & Yarnal, B. (2010a). Stakeholder perspectives on land-use strategies for adapting to climate-change-enhanced coastal hazards: Sarasota, Florida. *Applied Geography, 30*(4), 506–517.

Frazier, T. G., Wood, N., Yarnal, B., & Bauer, D. H. (2010b). Influence of potential sea level rise on societal vulnerability to hurricane storm-surge hazards, Sarasota County, Florida. *Applied Geography, 30*(4), 490–505.

Genovese, E., & Green, C. (2014). Assessment of storm surge damage to coastal settlements in Southeast Florida. *Journal of Risk Research, 18*(4), 407–427.

Gillis, J., & Barringer, F. (2012, November 18). As coasts rebuild and U.S. pays, repeatedly, the critics ask why. *New York Times*. Retrieved from http://www.nytimes.com/2012/11/19/science/earth/as-coasts-rebuild-and-us-pays-again-critics-stop-to-ask-why.html?pagewanted=all&_r=0

Keellings, D., & Waylen, P. (2014). Increased risk of heat waves in Florida: Characterizing changes in bivariate heat wave risk using extreme value analysis. *Applied Geography, 46*, 90–97.

Montz, B. E., & Tobin, G. A. (2011). Natural hazards: An evolving tradition in applied geography. *Applied Geography, 31*(1), 1–4.

Murley, J., Heimlich, B. N., & Bollman, N. (2008). *Florida's resilient coasts—A state policy framework for adaptation to climate change*. Fort Lauderdale: Florida Atlantic University Center for Urban and Environmental Solutions and National Commission on Energy Policy.

National Oceanic and Atmospheric Administration & the Environmental Protection Agency. (2011). *Achieving hazard and waterfront smart growth*. Retrieved from http://coastalsmartgrowth.noaa.gov/resilience.html

Natural Weather Service National Hurricane Center. (2014). *Sea, lake, and overland surges from Hurricanes (SLOSH)*. Retrieved from http://www.nhc.noaa.gov/surge/slosh.php#SDISPLAY

NatureServe. (2013). *Tools for coastal climate adaptation planning: A guide for selecting tools to assist with ecosystem-based climate planning*. Retrieved from https://connect.natureserve.org/sites/default/files/documents/EBM-ClimateToolsGuide-FINAL.pdf

Ruppert, T. (2011). Tools in the resilience toolbox, but are we willing to use them? *Ocean and Coastal Law Journal, 16*(2), 551–560.

Schapley, P. M., & Schwartz, L. (2014). Coastal hazard mitigation in Florida. In A. Farazmand (Ed.), *Handbook of crisis and emergency management* (pp. 771–789). Boca Raton: CRC Press.

Schwab, J. (2011). *Hazard mitigation: Integrating best practices into planning*. Retrieved from http://www.fema.gov/media-library/assets/documents/19261

Tobin, G. A. (1999). Sustainability and community resilience: The holy grail of hazards planning? *Global Environmental Change Part B: Environmental Hazards, 1*(1), 13–25.

Chapter 4
Hazards

Abstract In Florida, hazards can manifest themselves in a variety of forms, ranging from natural to man-made, and, at times, a combination of the two. Preparation for, and knowledge of, local hazards through mitigation methods and planning are of supreme importance to ensure a quick recovery from whatever problem comes knocking. By understanding the potential impacts through predictive methodologies and developing resilience efforts, administrators can increase their community's capacity to mitigate, respond and recover from anything that may impact their area. Education is powerful as Florida presents unique environmental issues, which affects recovery efforts. It is the responsibility of those involved to understand the nuances of their communities, building strong relationships with the citizens, and develop action plans to prepare for any potential hazard and reduce consequential impacts.

Keywords Hazards • All-hazards • Disaster preparedness • Assessment tools • Rural communities • Environmental issues • Social and economic issues • Action plans • Florida

Hazards can occur in many different forms and can affect communities in very different ways and become disasters after impact when vulnerabilities are exposed and society is interrupted (Kapucu and Özerdem 2013; McEntire 2005). Natural Hazards tend to be more common and cannot be prevented whereas technological hazards are caused by man and affect surrounding natural and non-natural environments alike and are almost always preventable (Kapucu and Özerdem 2013). Natural hazards include meteorological hazards, hydrological hazards, and geophysical hazards. Human made disasters are either technological in design or socially driven (See Fig. 3.1).

The difference between a hazard and a disaster is the human factor. Disasters are where hazards and human vulnerability collide, resulting in tragic loss of life, injuries, economic and social problems, as well as ecological problems (Kapucu 2012; Porwal et al. 2011). Natural hazards are more predictable than man-made disasters, especially with modern technological advancements. A common example of this is in hurricane predictions. Even if the precise direction can be difficult to determine, the general direction can be projected, allowing citizens ample time to prepare (Kapucu and Özerdem 2013) (Fig. 4.1).

© Springer International Publishing Switzerland 2015   43
F.I. Rivera, N. Kapucu, *Disaster Vulnerability, Hazards and Resilience*,
Environmental Hazards, DOI 10.1007/978-3-319-16453-3_4

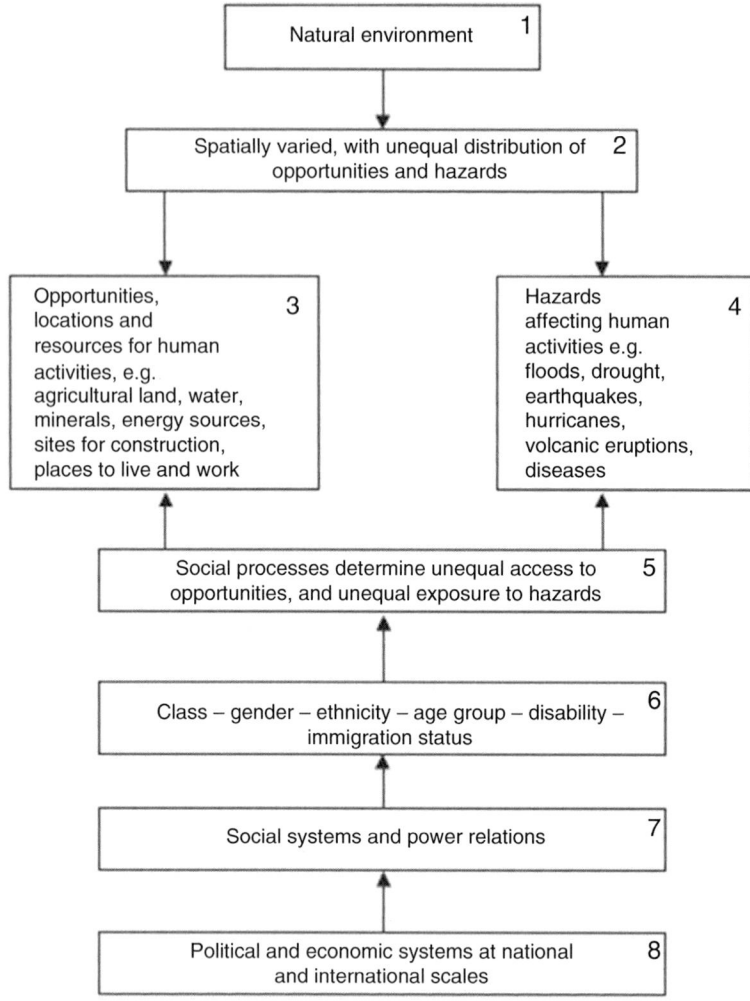

Fig. 4.1 Social causation of disasters (Wisner et al. 2004)

4.1   Definition of Hazard

Within many fields, there is a conversation regarding the importance of common terminology and providing a foundation for conversation through mutual under-standing. Hazards are defined as physical activities, phenomena, or human activities having the potential to cause injury, loss of life, damage to property, economic and social disruption, or environmental degradation (Kapucu and Özerdem 2013; Makoka and Kaplan 2005). The Florida Comprehensive Emergency Management Plan (FCEMP) incorporates an all-hazards approach, which holds the assumption of emergency support functions being similar despite the type of hazard (Florida Division of Emergency Management and State Emergency Response Team 2012). Hazards can run the gamut from man-made to natural to technological and are aided through the creation of comprehensive plans integrating all types verses individualistic protocols.

Differing slightly from the FCEMP, Lindell et al. (2007) categorized the principal hazards in the U.S. as meteorological, hydrological, geophysical, technological, and biological. Many of these hazards are typical of a certain region or landscape, but all areas are vulnerable to technological and biological threats. Emergency managers tend to specialize in mitigating the risk associated with a particular threat they are most likely to encounter; yet they must also prepare for an unlikely occurrence. For instance, a region prone to hurricanes should also have plans in place for radioactive spills and biological hazards since the high winds may damage industrial assets within the community.

In some situations, hazards can actually confound to create new ones. Lightning spawned by a common storm during Florida's summer months can set fire to wooded areas and endanger homes. Drought-like conditions present another factor for emergency managers to deal with, so they must work together with forestry departments to ensure the loss of woodlands and property are minimized. Lightning and droughts can be linked to meteorological hazards, which include severe storms, severe summer storms, tornadoes, and hurricanes (Kapucu and Özerdem 2013).

Florida, particularly Central Florida, is known for its vulnerability to tornadoes, wildfires, and hurricanes (Collins and Kapucu 2008; Kapucu 2008; Oxfam 2009). Rural communities are particularly susceptible to these hazards, especially in the U.S. Southeast (Oxfam 2009) and other rural areas in the world like the Asian-Pacific region (ESCAP/UNISDR 2012). In the following section we discuss some of the natural disaster hazards in Florida including wildfires, tornadoes, and an extensive discussion of hurricanes, as is one of the most prominent hazards in the region.

As for methods of assessment, Cutter (2001) noted several tools based on comparison of communities to geographic mapping. Florida had previously developed manuals utilizing Geographic Information Systems (GIS) based approaches to map out the state and identify vulnerable areas. By understanding the effects of coastal erosion, wind, storm surge, flooding, and other natural climate issues, local, state, and federal officials are better able to predict disaster prone communities and create proactive ways for improvement. The methodology chosen, to assess the area, is important as it "enables the user to examine what specific factors are most influential in producing the overall vulnerability" (Cutter 2001, p. 28).

Another method is analyzing the economic impact of a disaster (Kapucu and Özerdem 2013). For any hazardous situation, the physical impacts are considered property damage, loss of life, and other financially measurable variables. By determining the total amount a situation costs, the impact can be followed and used as predictive tools for future events. A drawback can be the initial investments needed to strengthen the area (National Academy of Science 2012). A tool to assist in this process is through HAZUS, developed by FEMA, to generate a probabilistic

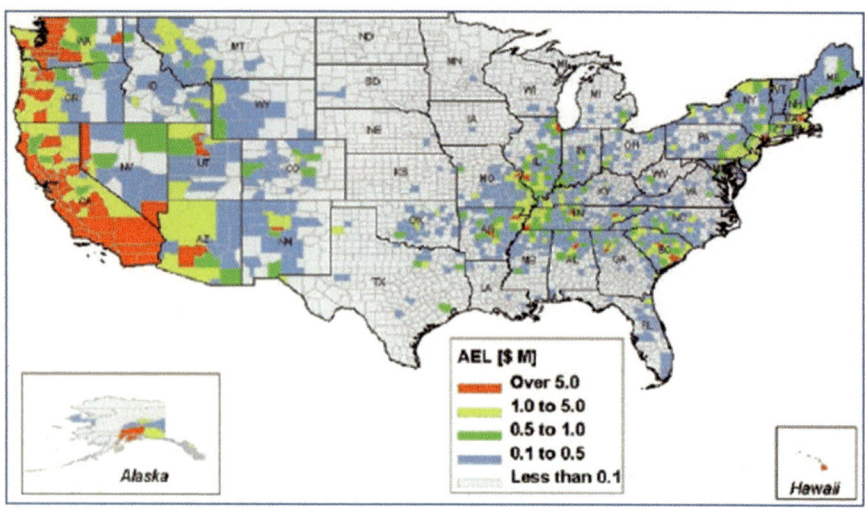

Fig. 4.2 Map representing example of HAZUS results (National Academy of Science 2012)

estimate of economic loss. An example of HAZUS results is seen in Fig. 4.2, which presents the seismic activity of the United States and the total value of loss determined by the exposure inventory.

4.2   Hazards in Florida

Due to Florida's unique climate, the state is vulnerable to a plethora of hazards including: "nuclear power plant accidents, hazardous materials incidents, mass communication failures, major power disruptions, oil spills, and critical infrastructure disruption/failure…terrorist attacks and mass migration events" (FDEM and SERT 2012, p. 8). Some of these hazards are created because of the high number of tourist, military, and government locations. Furthermore, the proximity of Florida to the Caribbean is a concern since the political environments are considered oppressed. It is crucial for managers to understand the nuances of their environments when it comes to "social, economic, historic, and political process that [may] impinge on a social group's ability to cope with contemporary hazards events" (Cutter 2001, p. 14).

The contextual paradigm can be generated through the inclusion of six thematic areas:

1. "Identification and mapping of the human occupancy of the hazard zone,
2. Identification of the full range of human adjustments to the hazard,
3. Study of how people perceive and estimate the occurrence of hazards,

4. Description of the processes whereby mitigation measures are adopted, including the social context within which that adoption takes place,
5. Identification of the optimal set of adjustments to hazards and their social consequences, and
6. Formulation of systems models to provide causal mechanisms for linking natural events and societal responses" (Cutter 2001, p. 5).

Although the context within paradigms is debated in regards to the inclusion of various aspects, critics agree there is, at least, a base to begin understanding hazards and cultural context is a highly influencing factor (Cutter 2001; Kapucu 2012).

4.3 Hurricanes

Florida is synonymous with hurricanes and it sort of an identity for the state (e.g. Hurricanes is the sports team for the University of Miami). Hurricanes, formed by the heating of the ocean, are a yearly occurrence for Floridians. These formations help release energy from the ocean and regulate temperatures (Kapucu and Özerdem 2013). Hurricanes are predictable to a point making early warning very effective, but they are also extremely powerful due to their high wind, possibility of tornadoes, torrential down pours, storm surges, and other hazardous outcomes that accompany a severe storm of this magnitude. Exposure to all of these events are inevitable when living in Florida so knowing the dangers and being proactive in the protection of our most vulnerable areas and people can minimize damages and human loss.

Its close proximity to the tropics and warm waters along the Atlantic Ocean, Gulf of Mexico, and Caribbean Sea puts Florida at a particular risk for hurricanes. In 2004 Florida experienced four major hurricanes (Charley, Frances, Jeanne and Ivan) resulting in the loss of 117 lives and more than $45 billion in damages (Newman 2004). Other associated hazards include storm surges, inland flooding, high winds, and tornadoes (FDEM 2012). This constant threat has created a very effective emergency management system that has been lauded as one of the best in the United States (Waugh 2006) and as a model for best emergency management practices (Demiroz et al. 2013). Although Florida has not experienced a major hurricane strike since 2004, it still remains the main focus of emergency management and response.

4.4 Tornados

Severe Storms can last for a matter of minutes, but bring high winds, lightning, large quantities of water, and create hazardous conditions on roads. Tornadoes, unlike hurricanes, are unpredictable and can occur during severe storms with potentially deadly consequences. Tornadoes are formed with the replacement of descending cold air with the rising hot air. Five out of 10 metropolitan areas prone to tornadoes reside in Florida, with 900 tornadoes forming in the U.S. annually (Kapucu and Özerdem 2013).

Tornadoes are a constant hazard in Florida and can occur in all seasons, but primarily during the spring (February-May) and summer (June-September) months. Summer tornadoes typically occur during the day and are due to strong sea breeze boundary coalitions or as a result of tropical cyclones. Spring tornadoes tend to be more powerful and deadly as they often strike in the middle of the night, which makes warning and escape less viable. Moreover, these powerful and deadly tornadoes can occur during the winter season months (October-January) (FDEM 2012). Of the five most deadliest Tornado outbreaks, two occurred in the Central Florida region, such as the 2007 Groundhogs Day tornado that killed 21 persons in Lake County and the seven tornadoes experienced in Orange, Osceola, Seminole, and Volusia Counties in February 22–23, 1998 resulting in the loss of 42 lives.

4.5   Wildfires

According to a report by the Florida Department of Agriculture and Consumer Services, Division of Forestry (2010) around 60 % of the land in Florida requires some type of natural fire or prescribed fire to maintain its ecosystem. Wildfires are common in North America (100,000 reported annually) and in Florida (5,550 reported annually). Several wildfires have caused significant damage in the area including the Mallory Swamp Fire in 2001, which burned more than 60,000 acres and was responsible for over $10 million timber loss, and the Bugaboo Scrub Fire in 2007, which forced the closure of three major highways with heavy smoke visible from Central Florida to Atlanta, Georgia (Florida Department of Agriculture and Consumer Services 2010). At any given time wildfires are burning in Florida. The hazard of wildfires are compounded by population growth accelerating the development of wild lands along with human behaviors being responsible for other 80 % of reported wildfires (FDOF 2010).

4.6   Lightening

Even though lightening is debated to be the deadliest, yet underrated, of natural hazards since it usually kills only one person at a time. However, there are an estimated 100 strikes per second and these electrical charges are one of the main reasons for ignited forest fires (Mileti 1999). Furthermore, this hazard is unique as its death toll consists of more than 80 % being males. Reasons for the statistic are given to the seasonality and time of the day of when the cloud-to-ground flashes occur (Cutter 2001).

4.7 Floods

Fluctuating waters in lakes or surface ponding occur seasonally. Dam failures happen when downstream water rises past a sustainable amount (Kapucu and Özerdem 2013). Floods are a costly and common natural hazard with 75 % of Presidential Disaster Declarations being ordered for flooding (Kapucu and Özerdem 2013). A 100-year flood plan is an assessment, not a prediction, of an area's flood incidents from the last century and can be utilized as a tool to help locate flood-prone areas. Storm Surge, most commonly seen with hurricanes, consists of seawater rising above the land in intense storms and can be the most damaging part with a capability of wiping out an urban area. Tsunamis, which usually occur with earthquakes, are extremely destructive and have a lot of issues due to the aftermath of the wave crashing into a coastal area.

4.8 Man-Made Hazards

Man-made disasters are accidents at best, terrorism at the worst. The human factor makes these hazards unpredictable, so the best course is to integrate everyday mitigation efforts as to not be caught off guard. These include regular and routine maintenance on machines and vehicles for transportation, back up plans and fail-safe measures, and proper planning and training to deal with terror threats.

Technological hazards are man-made and can include structural or engineering failure, transport, nuclear or environmental disaster, whereas social driven disasters include workforce violence, criminal violence, riots, stampedes and war (Kapucu and Özerdem 2013). These hazards are considered more preventable even with the potential to devastate an area. One of the greatest technological disasters to happen to the U.S. occurred in 2010 with the failure of a British Petroleum deep-sea oil drill. Not only did it kill 11 people with the initial blowout, it was one of the region's worst environmental disasters spilling over 4.9 million barrels of oil into the Gulf of Mexico causing adverse effects to the entire region (Kurtz 2013).

4.8.1 Hazards in Rural Communities

Citizens in rural communities encounter many challenges in the face of a hazard. Due to geographic, social, economic and political disconnect, the members of a community may find basic needs being inadequately addressed (Cross 2001; Tobin 1999). Moreover, recovery efforts can be stunted as many impacts of a hazard result in long-term rebuilding activities. Tobin (1999) proclaims the importance of cultural context when it comes to "returning a community to the status quo" or making "changes in the structure and thinking of society to accommodate hazards" (p. 15).

Bluntly spoken, Oliver-Smith (1996) denotes disasters and hazards as failures of society to adapt to the social and natural needs of the community. For example, Hurricane Andrew heavily impacted homeless individuals of downtown Miami area. These citizens comprised almost 10 % of the total metropolitan population (Cross 2001). For rural communities, these individuals are known to experience delays in response efforts due to physical distance, limited resources and just the lack of awareness to their situation.

By incorporating this knowledge in preparation, mitigation, response and recovery efforts, rural communities have a better chance at attaining the following characteristics of a resilient community:

- Lowered levels of risk to all members through reduced exposure to the geophysical event; reduced levels of vulnerability for all members of society;
- Ongoing planning for sustainability and resilience;
- High level of support from responsible agencies and political leaders;
- Incorporation of partnership and cooperation at different governmental levels;
- Strengthened networks for independent and interdependent segments of society; and
- Planning at the appropriate level (Tobin 1999, p. 17).

To attain the aforementioned characteristics, Oliver-Smith (1996) mentions the need to understand response behavior on a holistic level by incorporating behavioral, social, and political approaches. For rural communities, this is especially important as their inherent vulnerability greatly impacts their ability to prepare and recover.

4.9 Hazard Perception

Even though natural hazards can be predicted, they cannot be prevented. Therefore, the only effective way to be resilient is through mitigation, being prepared, having an effective response, and being able to recover (Kapucu et al. 2008; Kapucu and Özerdem 2013). By design, opinions on what constitutes a disaster and the effects of such are divergent (Mileti 1999). Basic perceptions may declare natural disasters to be those of greatest magnitude and largest affect verses technological hazards, emergencies and more. Mileti (1999) details the view of Dombrowsky who argued disasters incorporate a combination of intangible aspects. One cannot know the true consequences of society's behaviors and these actions can interact with technological and natural processes in unforeseen ways. This lack of predictable only adds to the issue of hazard perception and muddles the answer of "what is a disaster?"

When declaring a disaster, the FCEMP (2012) details the administrative process, which entails assumptions hinging on resource availability, allocation and responsibility. Furthermore, the procedure includes the need for state and federal support even if an event is considered locally concentrated. In addition, there is a necessity for an organized, comprehensive plan of action incorporating private, public and nonprofit organizations. Generated action items are not only useful for responding to a hazard, but also for mitigating the effects on communities in terms of perception. Cutter (2001) proclaimed one opinion in Florida being seen as a disaster prone state and has a "Wizard of Oz" effect in how society sees the climate verses the reality.

Regardless of the way comprehensive action plans are created, there is a commonality of response and recovery efforts including four beliefs: "(1) Technological and natural disasters are distinct; (2) The natural/technological distinction is not important; (3) Disasters as social constructions; and, (4) Human agency as the unifying factor" (Mileti 1999, p. 212). By understanding the beliefs and perceptions behind hazards, administrators increase their capability in improving a community's resilience especially since the long-term impacts can be difficult to foresee, as there are variables you cannot control for, such as population growth (Fig. 4.3).

Throughout the hazard event, resource attainment and distribution becomes critical. For Florida, the ability for emergency management officials to obtain necessary supplies and funds is dependent upon a multi-layered communication network. A visual representation of a potential process is seen in Fig. 4.4 where a plethora of administrators come to the table and assist in the decision-making process regarding necessary items. For example, if there is flooding on the coastal area of Florida, then the affected areas can come together with state officials, if applicable, and mitigate the impact through prioritization and allocation of responsibilities. Even though the state may provide funding, it could be up to local nonprofits to find housing or businesses to serve food and water to impacted individuals.

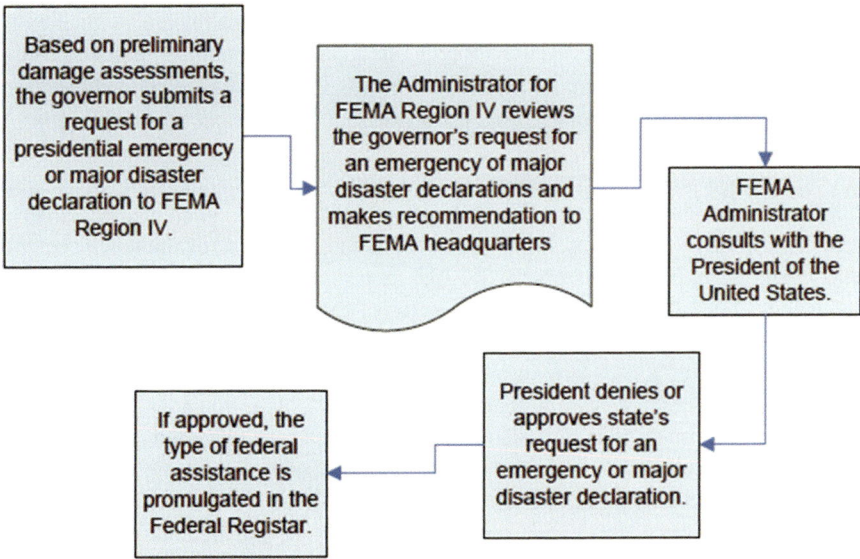

Fig. 4.3 Declaration process of Florida disasters (FCEMP 2012)

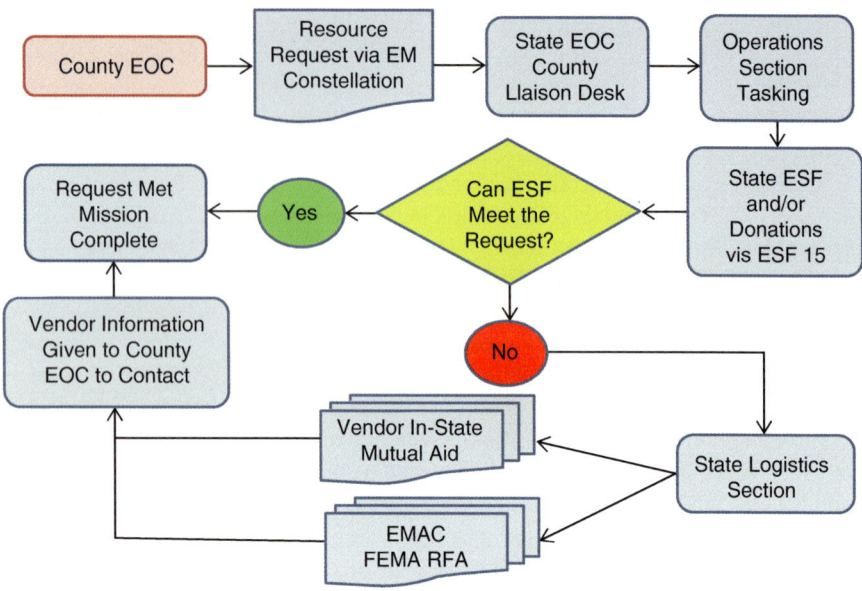

Fig. 4.4 Resource Allocation Model (FCEMP 2012)

By building positive networks, a community's resiliency increases. The Committee on Increasing National Resilience to Hazards and Disasters along with the Committee on Science, Engineering, and Public Policy (2001) found the following to be action items proven to foster resiliency: engagement of the citizenry in policymaking and planning, connecting infrastructure performance to resiliency goals, discussing the risks and educating the area, and incorporating sound land-use planning practices, building codes and standards into existing hazard mitigation plans. By incorporating the community, goals can be achieved as diverse groups unite for the common good.

We analyzed the focus group data in regards to the participant's experiences with different hazardous situations. Not surprisingly, hurricanes were mentioned as a hazard in all 7 focus groups. Yet, there was a shared concern of the sole focus on hurricane might be shortsighted, as one participant from rural Lake County indicated: "So, I hate to say it but the center of the state is neglected because the storm starts to de-intensify, you know, as it comes in. And Florida is only mainly thinking about hurricanes you know, it's the sudden tsunami or the sudden impact, it's the other all hazards events that Florida will be going, you know," "what was that?" This sentiment was also shared by a participant in the more urban Orange County who stated, "Well, and I think a lot of times that we focus, not us as a group, but people, focus only on hurricanes, not all hazards."

Participants were concerned about other types of hazards as well and the unique challenges they bring. For instance, a participant from coastal Brevard County indicated the potential for other hazards including terrorism, wild fires, and tornadoes.

I think it's also important to talk about hazards. We are one of the few places that are vulnerable to as many hazards as we are in the county, or in the state. Because of our tourism and our port, and our infrastructure, we are just as vulnerable to terrorism as the larger areas. We're costal, so we're long and we have hurricane vulnerability, we have tornado vulnerability, we have a lot of wild lands, so we've very vulnerable to wild fires. We're also very vulnerable to tornadoes. Um, pretty much any of the hazards that the state of Florida is vulnerable to, we've got it.

Other hazards in this County included dealing with a power plant, space launches (Kennedy Space Center launch pad is located in this County), and the potential damage to bridges that connect the shore areas to the inland.

Participants from all focus groups mentioned their experiences with tornadoes and wildfires, particularly in those counties that experienced the February 1998 tornadoes (Orange, Osceola, Seminole, and Volusia Counties) and the 2007 Groundhogs Day tornadoes (Lake County). There were other particular hazards mentioned in counties with rural areas. For instance the particularities of flooding, as one Seminole County participant stated: "I think that we really have a lot more flooding issues, with homes getting flooded. And from the flooding comes other things. You know, you get rat infestations, you get snakes, you get disease, you get mold. We've had a huge issue with the flooding and the mold that comes with it."

Overall, hurricanes, tornadoes, floods, and wildfires were identified as recurring hazards communities in the seven Central Florida counties continue to experience and prepare for. Others hazards included terrorism, mold, and damaged bridges and are, also, experienced in other places within the U.S. and the world. In the next section, we provide a short content analysis of news articles during the 2014 hurricane season.

4.9.1 Content Analysis for 2014 Hurricane Season

To find the latest news related to the 2014 Florida Hurricane Season, searches were conducted utilizing the library resources at the University of Central Florida, along with search engines such as Google and individualized searches on the following news venues: New York Times, Miami Herald, and Orlando Sentinel. The results were refined to showcase reports from June 2014 to August 2014 with the majority of information related to preparation aspects and insurance awareness. We summarized the findings for articles focusing on hurricane preparedness, resilience, vulnerability and hurricane hazards (See Appendix C for a detailed description of each news article).

4.9.2 Hurricane Preparedness, Resilience, Vulnerability, and Hurricane Hazards

There were several reports showcasing hurricane preparedness including hearings by the U.S. Senate and Public Works Subcommittee on Clean Air and Nuclear Safety that emphasized the importance and influence of risk insurance as a preparedness strategy in the face on recurring national catastrophes. In addition, there were reports on the importance of preparedness activities to prevent environmental harm by the Florida Department of Environmental Protection, including mock disaster exercises, citizen's awareness of insurance policies, and a call for the governor to Floridians to plan and prepare for the hurricane season. Regarding resilience news articles emphasized on the need for action from the public and private sectors including efforts by the U.S. Congress to increase legislative efforts to strengthen building standards and disaster savings effort for future disaster events and some research efforts to create hurricane resilient coastlines in the U.S. The emphasis for vulnerability was on poor communities that highlighted the conditions by which weather events affect the poor the hardest. Finally, news articles with regards to hurricane hazards highlighted potential problematic issues such as hurricane shutters with no escape latch and the ongoing sea-level rise threat experienced in the Miami-Dade area.

4.9.3 Conclusion

The exposure of Florida to multiple hazards (either technological or natural) required an all hazards approach to disaster preparedness. These all hazards approach benefits from an array of measurement tools that can assess the geographical, social, and economic hazard impacts. By understanding these impacts administrators can increase their community's capacity to mitigate, respond and recover from anything that may impact their area. In addition, these tools can are capable of increasing knowledge, response behavior, hazard perceptions. Understanding these factors can facilitate better disaster declarations and improve citizen behavior. For some in rural communities, the all hazards approach is welcomed, but there is still concern that policies and emphasis is still primarily concern exclusively with hurricanes, which tend to be less hazardous in rural-inland communities. In all, Florida presents unique environmental, social, and economic issues, which affects recovery efforts. It is the responsibility of those involved to understand the nuance of their communities, build strong relationships with citizen, and develop action plans to prepare for any potential hazard and reduce consequential impacts.

References

Collins, M. L., & Kapucu, N. (2008). Early warning systems and disaster preparedness and response in local government. *Disaster Prevention and Management, 17*(5), 587–600.

Committee on Increasing National Resilience to Hazards and Disasters & Committee on Science, Engineering, and Public Policy. (2001). *Disaster resilience: A national imperative.* Washington, DC: The National Academies.

Cross, J. A. (2001). Megacities and small towns: Different perspectives on hazard vulnerability. *Global Environmental Change Part B: Environmental Hazards, 3*(2), 63–80.

Cutter, S. (Ed.). (2001). *American hazardscapes: The regionalization of hazards and disasters.* Washington, DC: John Henry Press.

Demiroz, F., Kapucu, N., & Dodson, R. (2013). Community capacity and interorganizational networks for disaster resilience: Comparison of rural and urban counties in Central Florida. In N. Kapucu, C. Hawkins, & F. Rivera (Eds.), *Disaster resiliency: Interdisciplinary perspectives* (pp. 334–354). New York: Routledge.

Economic and Social Commission for Asia and the Pacific/International Strategy for Disaster Reduction (ESCAP/UNISDR). (2012). *Asia Pacific disaster report 2012: Reducing vulnerability and exposure to disasters*. Retrieved from http://www.unescap.org/resources/asia-pacific-disaster-report-2012-reducing-vulnerability-and-exposure-disasters

Florida Comprehensive Emergency Management Plan (FCEMP). (2012). *The state of Florida Comprehensive Emergency Management Plan 2012*. Retrieved from http://floridadisaster.org/documents/CEMP/2012/2012%20State%20CEMP%20Basic%20Plan%20-%20Final.pdf

Florida Department of Agriculture and Consumer Services (FDOF). (2010). *Wildfire risk reduction in Florida: Home, neighborhood and community best practices*. Retrieved from http://freshfromflorida.s3.amazonaws.com/Wildfire_Risk_Reduction_in_FL.pdf

Florida Division of Emergency Management (FDEM), & State Emergency Response Team (SERT). (2012). *Comprehensive emergency management plan*. Retrieved from http://floridadisaster.org/documents/CEMP/2012/2012%20State%20CEMP%20Basic%20Plan%20-%20Final.pdf

Kapucu, N. (2008). Collaborative emergency management: Better community organizing, better public preparedness and response. *Disasters: The Journal of Disaster Studies, Policy, and Management, 32*(2), 239–262.

Kapucu, N. (2012). Disaster resilience and adaptive capacity in Central Florida, US, and in Eastern Marmara Region, Turkey. *Journal of Comparative Policy Analysis: Research & Practice, 14*(3), 202–216.

Kapucu, N., & Özerdem, A. (2013). *Managing emergencies and crises*. Boston: Jones & Bartlett Publishers.

Kapucu, N., Berman, E., & Wang, S. (2008). Emergency information management and public disaster preparedness: Lessons from the 2004 Florida Hurricane season. *International Journal of Mass Emergencies and Disasters, 26*(3), 169–197.

Kurtz, R. (2013). Oil spill causation and the deepwater horizon spill. *Review of Policy Research, 30*(4), 366–380.

Lindell, M. K., Prater, C. S., & Perry, R. W. (2007). *Introduction to emergency management*. Hoboken: Wiley.

Makoka, D., & Kaplan, M. (2005). *Poverty and vulnerability. An interdisciplinary approach*. Bonn: Universitat Bonn.

McEntire, D. A. (2005). Why vulnerability matters: Illustrating the need for a modified disaster reduction concept. *Disaster Prevention and Management, 14*(2), 206–222.

Mileti, D. S. (1999). *Disasters by design: A reassessment of natural hazards in the United States*. Washington, DC: Joseph Henry Press.

National Academy of Sciences (NAS). (2012). *Disaster resilience: A national imperative*. Washington, DC: The National Academies Press.

Newman, J. (2004). Numbers tell story of season's destruction. *Orlando Sentinel*, pp. A1, A10.

Oliver-Smith, A. (1996). Anthropological research on hazards and disasters. *Annual Review of Anthropology, 25*, 303–328.

Oxfam. (2009). *Exposed: Social vulnerability and climate change in the US Southeast*. Washington, DC: Oxfam.

Porwal, M. C., Padalia, H., & Roy, P. S. (2011). Impact of tsunami on the forest and biodiversity richness in Nicobar Islands (Andaman and Nicobar Islands), India. *Biodiversity and Conservation, 21*(5), 1267–1287. doi:10.1007/s10531-011-0214-x.

Tobin, G. A. (1999). Sustainability and community resilience: The holy grail of hazards planning? *Global Environmental Change Part B: Environmental Hazards, 1*(1), 13–25.

Waugh, W. L., Jr. (Ed.). (2006). Shelter from the storm: Repairing the national emergency management system after Hurricane Katrina. *Special issue of The Annals of the American Academy of Political and Social Science, 604*, 256–272.

Wisner, B., Blaikie, P., Cannon, T., & Davis, I. (2004). *At risk: Natural hazards, people's vulnerability and disasters*. New York: Routledge.

Chapter 5
Vulnerability

Abstract In this chapter we discussed the concept of vulnerability. This will be achieved through a detailed discussion of social, economic, geographical, and political attributes of Florida communities and regions that shape vulnerabilities to disasters. The chapter also provides a detailed analysis of the perceptions of vulnerabilities among several emergency management communities including a discussion on poverty, homelessness, the elderly, and other vulnerabilities. Furthermore, we discuss several tools for the identification of vulnerability. Overall, physical and social vulnerabilities will change and administrators must keep this aspect in the forefront. Battling the complacency within a community cannot be done if their officials also become complacent. It is imperative for emergency management administration to stay up to date on legislation, policies and procedures, which will affect their citizenry.

Keywords Vulnerability • Risk management • Social and economic issues • Complacency • Disaster vulnerability identification tools • Mistrust • Capacity building • Florida

Emergency Management is a relatively new and interdisciplinary profession. Past experiences have taught public administrators and policymakers the importance of looking holistically at all phases of EM: preparation, mitigation, response and recovery (Waugh 1994). With each new disaster and hazard, the weaknesses of an area come to the service (Donner and Rodriguez 2008). It becomes the responsibility of the entire community to increase their resiliency efforts and decrease vulnerability (Committee on Increasing National Resilience to Hazards and Disasters, Committee on Science, Engineering, and Public Policy, & The National Academies 2012). To do so, managers must incorporate and analyze multiple layers, such as social, economic, political and geographic attributes, as each community is a unique amalgamation. For Florida, vulnerabilities are surfacing due to ever-changing climate and demographics resulting in at-risk cultural groups like the impoverished, homeless, elderly and more.

One of the most important ways to reduce the effects of a disaster and increase the resilience of a community is to closely examine/identify all the vulnerabilities of the community with regards to various hazards, build local capacity, and build

© Springer International Publishing Switzerland 2015   57
F.I. Rivera, N. Kapucu, *Disaster Vulnerability, Hazards and Resilience*,
Environmental Hazards, DOI 10.1007/978-3-319-16453-3_5

partnerships in developing 'culture of preparedness' and 'culture of prevention' (Kapucu 2008; Wisner 2009). This includes looking at the geography of the community, infrastructure, demography of its citizens, as well as historical events. An example of such an analysis was conducted by Randall Parkinson for the city of Satellite Beach, Florida in response to the threat of sea level rise from climate change (Parkinson and McCue 2011). In this analysis, models were used to predict the vulnerability of the city and indicate at-risk areas, or those where the worst impacts would be felt, and determine where the citizens could retreat. Models are some of the most useful tools for disaster resilience, as they allow managers to go through variable hazardous situations given certain parameters, and see what the aftermath would look like. This type of foresight enables managers to make appropriate decisions regarding mitigation efforts, which will in turn make communities more resilient.

In this chapter, we examine vulnerability of rural communities from the current literature and policy documents and provide some examples from Florida. We also provide perceptions of vulnerability from several communities in Florida.

5.1 Disaster Vulnerability

As stated in the introductory chapter, we define disaster vulnerability as "a concept that denotes a social practice in which a certain unit (a subject, a group, or any kind of system) is placed at the center of a complex analysis of injury" (Kusenbach and Christmann 2013, p. 64). We argue to truly understand vulnerability attention must be paid to important issues such as the social construction of disaster vulnerability. These concerns include: the cultural context of human perceptions and interpretations (particularly the perceptions of those in the front lines of disaster emergency management and response); the view of vulnerability as negative and resources as positive while often overlooking the unintended side effects of vulnerabilities and resources; and the complexities of time which alter the meaning of vulnerabilities before, during, and after a disaster situation.

Indeed, it is important to discuss the encompassing definition of 'community' in general (Committee on Increasing National Resilience to Hazards and Disasters, Committee on Science, Engineering, and Public Policy, & The National Academies 2012). For many, a community is a geographically defined term or focused on one cultural group who share similar characteristics. This discussion utilizes the term in its broadest sense to become an umbrella to all groups found within Florida, which includes any social, economic, political, or geographic boundary.

Discussing communities considered vulnerable means scholars are analyzing a community's ability to cope, prepare and recover from impacts of natural hazards (Donner and Rodriguez 2008; Waugh 1994). This definition incorporates a more sociological view and stems from the premise of vulnerability incorporating characteristics or social attributes (Cutter and Emrich 2006). For example, those deemed impoverished, such as the homeless, are considered a vulnerable community due to lack of resources to prepare, mitigate, respond and recover from a disaster.

By focusing on place-based characteristics, communities become unique social entities whose vulnerabilities vary in their creation and development (Myers et al. 2008). This provides another way to perceive vulnerability within its relationship with a hazard. A hazard is seen as a potentially damaging event with the capability to cause loss of life, property damage, negative socio-economic impacts, and environmental degradation. The affect is considered external.

Vulnerabilities, deemed an internal factor, can increase or decrease the hazard's impact (Kapucu and Özerdem 2013). Furthermore, vulnerabilities can be grouped into physical, "the culmination of human vulnerability, agricultural vulnerability, and structural vulnerability," (Kapucu and Özerdem 2013, p. 25) and social, "lack of access to resources, including information and knowledge; limited access to political power and representation; certain beliefs and customs; weak buildings or individuals; and infrastructure and lifelines" (p. 25).

5.2 Disaster Vulnerability in Florida

As previously mentioned, Florida holds unique overlapping ecological and social challenges affecting the field of emergency management (see Chap. 3). For example, a growing concern in the coastal area is the increase in the elderly population (Donner and Rodriguez 2008). Over the years, the sprawling into the coastal regions has led to more land development in already vulnerable areas, because of aspects like coastal erosion, along with an increase in the average age of the residents, which adds another layer to risk management when it comes to arenas such as evacuation concerns (Cutter and Emrich 2006; Mittler 1997). Granted, there is a challenge to administrators typecasting certain cultural groups as vulnerable, such as the elderly (Elmore and Brown 2007). Disaster management is a dynamic field where knowledge of a community is the only way to truly understand the capabilities of its members (Kapucu and Liou 2014).

Indeed, focus groups participants shared their views on whom they considered to be vulnerable and the elderly (along with the poor and special needs populations) were identified as particularly vulnerable to disasters. For instance, when asked whom they considered to be the most vulnerable to disasters in their respective counties participants from our focus groups stated:

"But a lot of elderly in uh, south Brevard, and also in north Brevard, and that's a big rural area up there in Mims, um that's a challenge to get rescue people there and to respond to them." (Brevard County).

"…but you know, when you get, the way that our demographics have shifted, the mobile home park folks are typically the older ones. They're the ones that are in their mid-70s to 90s. Um, so they're financial constrained, they're mobility challenged and different things like that" (Lake County).

"Um, a lot of the older people, um they, including my parents even, you know their idea of what they may need in case of an emergency, gas for their cars, uh, you know get things on order, call your family up if there's a disaster coming through or anything like that" (Orange County).

"...the elderly, the disabled, people who don't have any family and that type of thing" (Seminole County).

When evaluating vulnerable cultural groups, aside from the elderly and homeless, Florida administrators must take language barriers into account. Considered linguistic capital, there are a growing percentage of individuals whose first language is not English (Donner and Rodriguez 2008). Miscommunications can become detrimental when generating informational material or informing the public when a disaster is coming, occurring or passing (Kapucu et al. 2013a, b). To exemplify, among the focus participant's language and cultural issues were brought up in discussing vulnerable populations:

"...and then you run into the cultural issues where, uh, there may be communications issues, that inhibit them to understand what's being presented to them, and then too, having the capabilities to request assistance" (Orange County).

5.3 Disaster Vulnerability in Rural Communities

Communication and educational attempts can also become hindered when taking rural communities into account. Urban, or rural, areas are seen as those with limited resources and incorporate concerns due to their remote nature (Cutter and Emrich 2006; Donner and Rodriguez 2008; Prelog and Miller 2013). In addition, these areas possess concerns based on their demographic breakdown in terms of ethnicity and gender (Emrich and Cutter 2011; Prelog and Miller 2013). In addition, citizens in lower socioeconomic groups are more prone to housing damages in times like hurricane season due to the inability to properly prepare their homes, resulting in a cyclical effect of social vulnerability from the incapability to protect and then rebuild homes and communities (Cutter and Emrich 2006; Myers et al. 2008; Prelog and Miller 2013). Outside of the socio-demographic issues discussed before some of the major concerns from our focus group participants were physical and geographical issues, particularly mobile home parks and physically isolated communities. For instance,

"We're trying to push it out more to the mobile homes because they're the most vulnerable" (Lake County). "For us when you talk about vulnerability, a lot of it is geographical. Uh, it has to do with elevation and uh the same areas that are susceptible to flooding in some cases, in portions of the county like Poinciana and others, that same geography creates problems for us in that the sewer system can be inundated by those waters and create problems at all of our facilities so, uh, geography plays a good portion into what – I mean you have special needs persons and a lot of people living in mobile home parks and things of that nature that are susceptible to the various uh, things that may happen" (Osceola County). Furthermore, "for us in emergency management, we look to people that live in mobile homes, or manufactured homes" (Seminole County).

Much like the elderly population, rural communities are seen as susceptible to disasters and hazards with an assumption of citizens unable to recover as quickly as those in urban areas (Durant 2011). Cultural stereotypes can have a profound impact on the phases of EM due to the perception of vulnerability. It is crucial for administrators to be aware of their misconceptions to prevent the continuation of biased practices (Durant 2011; Prelog and Miller 2013).

Moreover, it is important for public officials to educate their population, along with other local, state, and federal officials, on the community's abilities to respond, recover and prepare citizenry as much as they can (Kapucu et al. 2013a, b; Waugh 1994). Strong social networks are critical for efficient and effective recovery.

Evaluation methods can consist of three social and ecological elements: the individual level, the community level, and the presence of vulnerabilities (Prelog and Miller 2013). Within the analytical period, administrators can discuss the weaknesses and strengths of each sphere. One strength Prelog and Miller (2013) point out is the community's connection during disaster response where the members are so accustomed to having to pull together that it becomes more natural to assist others verses urban areas. Planning designs and directive tools can visually convey this information, such as the one created by Stripling (2013) in the Disaster Planning Handbook (See Fig. 5.1).

Systems	Significant Characteristics	Operational Risks
Political	•	•
Response Capacity	•	•
Economic		•
Social	•	•
Information	•	•
Infrastructure	•	•
Physical Environment	•	•
Time	•	•

Fig. 5.1 Community systems impact: environmental frame (Stripling 2013)

5.4   Disaster Vulnerability Identification

In determining potentially vulnerable areas, it is imperative to take into account a multitude of variables (Tate 2012; Zahran et al. 2008). Mostly seen in a quantitative presentation, vulnerability assessments "identify the processes that produce vulnerability and associated variables that can be used to measure differential hazard susceptibility" (Tate 2012, p. 327). These can be observed in factors related to climate and cultural groups along with being linked to more of a theoretical overview verses practical or a culmination of all the above (Lindell 2013) (see Fig. 5.2 for suggested indicators).

A connected model to assist in the disaster impact is an adapted visual (See Fig. 5.3) from Lindell (2013) who distinguished effects into three categories of pre-impact, trans-impact and post-impact. Within each temporal period, there exist sub-sections related to the four phases of emergency management, which are found to be overlapping in nature and are incapable of being mutually exclusive. Additionally, there is further definition by the affinity of being a vulnerability, exposure or impact.

On the other hand, many researchers look to the ecological aspect and have created predictive models to assess climate variability like droughts, hurricanes, sea level rise and floods (Emrich and Cutter 2011). The Social Vulnerability Index

Stage	Description	Example options
Conceptual framework	Vulnerability dimensions to include	Access to resources, demographic structure, evacuation, institutional
Structural design	Organization of indicators within the index	Deductive, hierarchical, inductive
Analysis scale	Geographic aggregation level of indicators	US county, census enumeration unit, neighborhood, raster cell size
Indicator selection	Proxy variables for dimensions	Income, education, age, ethnicity, gender, occupation, disability
Measurement error	Accuracy and precision of the demographic data	Census undercounts, reported margin of error
Transformation	Indicator representation	Counts, proportions, density
Normalization	Standardization to common measurement units	Ordinal, linear scaling (min–max, maximum value), z-scores
Data reduction	Reduction of large correlated indicator set to a smaller set	Factor analysis
Factor retention	How many principal components to retain?	Scree plot, Kaiser criterion, parallel analysis
Weighting	Relative degree of indicator importance	Equal, expert, data envelopment analysis, budget allocation, analytic hierarchy process
Aggregation	Combination of normalized indicators to the final index	Additive, geometric, multi-criteria analysis

Fig. 5.2 Social vulnerability suggested index (Tate 2012)

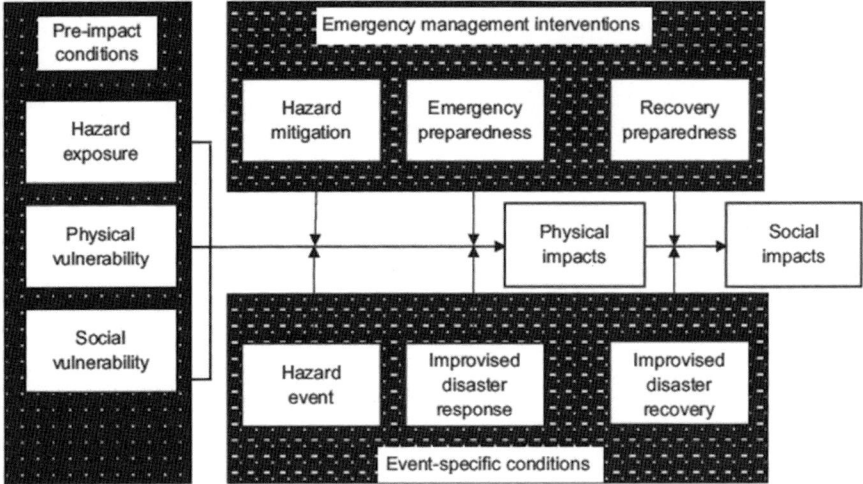

Fig. 5.3 Adapted disaster impact model (Lindell 2013)

Socioeconomic or demographic category	Variable description
Race/ethnicity	% black population, % Native American Indian population, % Asian or Pacific Islander population, % Hispanic population, % recent international migration
Wealth	Per capita income, % households earning more than $100,000 (U.S. dollars) in 2000, % living in poverty, median dollar value of owner-occupied housing units, median gross rent ($) for renter-occupied housing units
Housing type and tenure	% housing units that are mobile homes, no. housing units per square mile, % rural farm population, % urban population
Education and employment	% population over 25 yr old with less than 12 yr of education, % of civilian labor force unemployed, % civilian labor force participation, % female participation in civilian labor force, % employed in primary industry (farming, fishing, mining, forestry), % employed in transportation, communications, and other public utilities, % employment in service occupations
Age, gender, and health	% population under 5 yr old, % population 65 yr or older, average number of people per household, Social Security recipients, nursing home residents per capita, number of physicians per 100 000 population, number of hospitals per capita
Gender and family structure	% female population, % female-headed households, % renter-occupied housing units, % international migration

Fig. 5.4 Categorical grouping of variables used in the creation of the Social Vulnerability Index (Emrich and Cutter 2011)

(SVI) is one of the tools used to examine an area for the intersection of social vulnerabilities to the environmental hazards and produce a quantitative result (Cutter and Emrich 2006; Tate 2012) (see Fig. 5.4 for categories within the SVI and Fig. 5.5 for mapping results).

Another tool used is Geographic Information Systems (GIS), which is a mapping instrument to visually ascertain disaster-prone areas (Kapucu and Özerdem 2013). Some officials used GIS technology and discovered disconnects between rural areas and their closest emergency responders (i.e. fire and police). When a disaster occurs, there will be major time discrepancies between responder's abilities to get to the rural areas and assist those who need attention (Cutter and Emrich 2006).

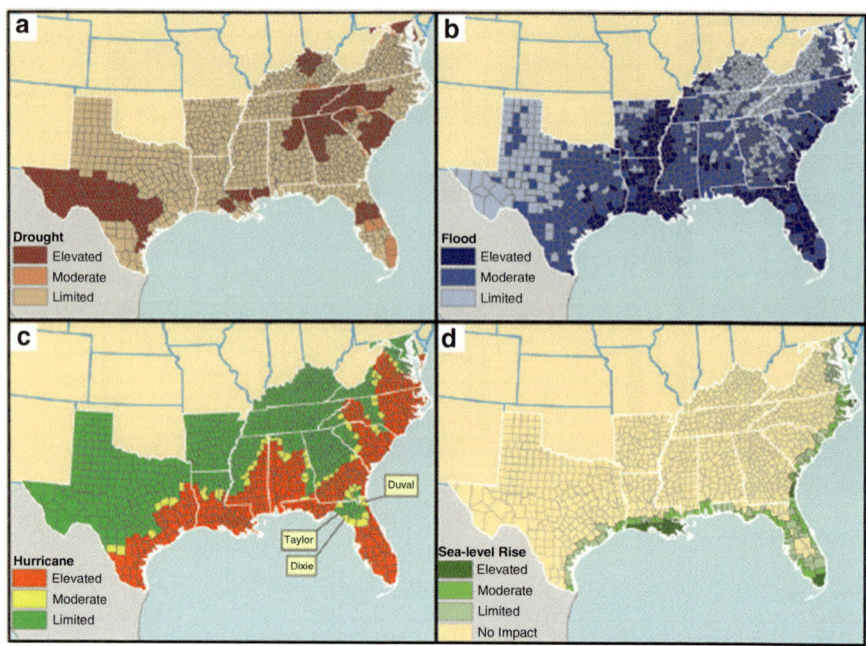

Fig. 5.5 Social vulnerability of the southeastern United States for drought, flood, hurricane and sea-level rise (Emrich and Cutter 2011)

In conjunction with the SVI and GIS, the University of South Carolina collected data on human and material costs related to disasters within the U.S. in a system called SHELDUS (Prelog and Miller 2013; Zahran et al. 2008). Considered the most comprehensive accumulation of response-related information, SHELDUS allows administrators to understand a longitudinal overview of impact as the program collected over 50 years worth of information.

Whichever tool or inventory is chosen, it is crucial for each state to assess their vulnerable areas. FEMA released a mandate for vulnerability assessments to be done and provided grant funding so state budgets did not have to feel the full affect (Kapucu and Özerdem 2013). The next step is incorporating the knowledge into preparation and mitigation strategies to develop less vulnerable communities.

Tierney (2012) aptly states that, "hazards and disasters do not respect borders" (p. 343). Disasters are a double-edged sword in a sense their impact is mostly seen as negative; however, they provide opportunities for a community to understand where their vulnerabilities lie and allow administrators the chance to enhance growth and development (Kapucu and Liou 2014). A recommendation by Elmore and Brown (2007) is to incorporate potentially affected populations in the education and decision-making process. More specifically, the elderly demographic is typecast as being vulnerable in every disaster or hazard; yet, this group is a wealth of information as their age provides wisdom of life experiences, which can be useful in each phase of emergency management.

Some generic steps to capacity building or resiliency include: "(1) build on local community knowledge of disasters; (2) prepare and deliver effective warning messages (news that will reach all diverse populations, including blind, deaf, and minority individuals); (3) develop shelters; (4) anticipate the need for translators, child care, medical equipment, and other resources; and (5) avoid wasting resources on unnecessary supplies due to bias" (Kapucu and Özerdem 2013, p. 28). Although these steps are proactive in nature, they can be easily translated into reactive methods as they focus on basic human needs. One arena is through hardening the system (Abi-Samra and Willis 2013). Hardening is the process of local officials ascertaining vulnerable utility systems to strengthen. The Florida Public Service Commission recently began hardening efforts in response to the Atlantic hurricane season of 2004–2005.

5.5 Disaster Reduction Strategies

A major issue to address when attempting to create vulnerability, or risk, reduction strategies is the complacency or fear many citizens feel when facing natural hazards. A prime example is the coastal regions of Florida where oceanfront properties are sought after even though they are a high area of concern when disasters strike (Donner and Rodriguez 2008). For many neighborhoods, hurricanes can result in evacuation requests. These appeals can be made in vain as some residents refuse to leave their homes out of complacency or fear. For undocumented residents, a request to evacuate is dismissed for fear of being deported back to their home countries (Donner and Rodriguez 2008; Waugh and Liu 2014).

Some scholars link the existence of complacency and fear to the aspect of trust. In the conversation of capacity building, a responsibility is placed upon administrators and officials to generate and maintain trustworthy relationships. Recent studies show racial and ethnic minorities hold a higher distrust of their local, state and federal representatives because of previous issues where promises were broken or relationships were taken advantage of (Donner and Rodriguez 2008). "Trust is a critical factor government officials should invest their time and resources in developing prior to a disaster as it enhances recovery processes and creates resiliency" (Kapucu and Liou 2014, p. 9). The members of a community are valuable resources, during a hazard or disaster, as they bring an aspect of volunteerism to the phases of EM. If a community consists of fragmented networks, then there is a decreased desire for members to assist in times of need (Waugh and Liu 2014).

Looking to the environmental factors, vulnerability reduction is seen through administrators who utilize results of risk identifiers, like the SVI, into their strategies for change. Mapping out the areas needing improvement can assist in the preparation efforts for future disasters. Knowing what Florida is prone too aids in the strategic planning for the state to prepare their citizenry to face them. For instance, outside of droughts, floods, sea-level rise and hurricanes, Florida is considered one of the top five vulnerable states for tornadoes, which is said to be the most violent type of storm to endure (Kapucu and Özerdem 2013). Not only is Florida one of the susceptible states, but also tornadoes are predicted to occur after midnight, which increases the predicted death toll as many people are asleep and are more at risk (Florida Disaster.org 2014). If local officials are able to evaluate their building codes, for example, then they can predict where physical structures may receive damage and can prepare to either rebuild or prevent the destruction (Kapucu and Liou 2014). "Disaster vulnerabilities can be reduced through pre-event activities, such as hazards and vulnerability assessments; land-use regulations; building code development, adoption, and enforcement; warning systems; and education and training programs" (Tierney 2012, p. 344).

For many administrators, there is a push for a proactive approach to disaster mitigation and risk management (Deconcini and Tompkins 2014). Recent actions include the Resilient Communities for America Agreement along with the President's Climate Action Plan. Each document outlines a push for communities to place an importance on anticipating disasters and hazards and building resiliency. Another reduction method is through a comprehensive planning approach where each state looks to their vision for land use and details out a way to "provide public benefits or present hazards to public welfare, including wetlands, productive coastal waters, wildlife habitat, scenic areas, historic and archeological resources, aquifer recharge areas, prime agricultural soils, floodplains, and other areas exposed to natural hazards" (Deyle et al. 2008, p. 350). For Florida, the creation of a comprehensive management plan began as early as 1990 and was useful in the response to Hurricane Andrew, but it was lacking during the implementation phases, as the state did not acknowledge every part of the document (Mittler 1997).

Similar to most legislative and planning efforts, it is the charge of administration to interpret policies while guiding the implementation process (Deyle et al. 2008). It is a fragile task as how managers enact their understanding can make or break a community. If there is a majorly negative impact, then trust can disintegrate causing issues for future actions. "Thus, effective local implementation of state goals is determined by the quality of the plan policies adopted by local governments, how local officials translate those policies into development controls and other growth management measures, and how they enforce those measures" (Deyle et al. 2008, p. 350). If the administration believes there are barriers and limitations, then it can impede a community's ability to move forward. Conversely, if there is a belief of no boundaries, then there is a chance for a similar impact.

5.5.1 Conclusion

It is important to recognize risk management is an ever-changing process due to changes in social, economic, geographic and political arenas. Social vulnerabilities will change and administrators must keep this aspect in the forefront. Battling the complacency within a community cannot be done if their officials also become complacent. Additionally, one must be aware of the acts perpetuating growth and those causing barriers. Fortunately, there are several disaster vulnerability identification tools that can be useful in identifying vulnerabilities within a community, which can lead to better planning designs and directive tools. Still, issues of public complacency and mistrust in governmental officials remain. Some ways to circumvent these issues include risk identifier tools and community planning efforts. In all, it is imperative for emergency management administration to stay up to date on legislation, policies and procedures, which will affect their citizenry.

References

Abi-Samra, N., & Willis, L. (2013). Hardening the system. *Transmission & Distribution World, 65*(2). Retrieved from http://tdworld.com/vegetation-management/hardening-system

Committee on Increasing National Resilience to Hazards and Disasters, Committee on Science, Engineering, and Public Policy, & The National Academies. (2012). *Disaster resilience: A national imperative*. Washington, DC: National Academies Press.

Cutter, S. L., & Emrich, C. T. (2006). Moral hazard, social catastrophe: The changing face of vulnerability along the hurricane coasts. *The Annals of the American Academy of Political and Social Science, 604*(1), 102–112.

DeConcini, C., & Tompkins, F. (2014). *Sea-level rise and its impact on Florida*. World Resources Institute. Retrieved from http://www.wri.org/profile/christina-deconcini

Deyle, R. E., Chapin, T. S., & Baker, E. J. (2008). The proof of the planning is in the platting: An evaluation of Florida's hurricane exposure mitigation planning mandate. *Journal of the American Planning Association, 74*(3), 349–370.

Donner, W., & Rodríguez, H. (2008). Population composition, migration and inequality: The influence of demographic changes on disaster risk and vulnerability. *Social Forces, 87*(2), 1089–1114.

Durant, T. J. (2011). The utility of vulnerability and social capital theories in studying the impact of Hurricane Katrina on the elderly. *Journal of Family Issues, 32*(10), 1285–1302.

Elmore, D. L., & Brown, L. M. (2007). Emergency preparedness and response: Health and social policy implications for older adults. *Generations, 31*(4), 66–74.

Emrich, C. T., & Cutter, S. L. (2011). Social vulnerability to climate-sensitive hazards in the Southern United States. *Weather Climate & Society, 3*(3), 193–208.

Floridadisaster.org. (2014). *Tornadoes*. Retrieved from http://www.floridadisaster.org/EMTOOLS/Severe/tornadoes.html

Kapucu, N. (2008). Culture of preparedness: Household disaster preparedness. *Disaster Prevention and Management: An International Journal, 17*(4), 526–535.

Kapucu, N., & Liou, K. T. (2014). Disasters and development: Investigating an integrated framework. In N. Kapucu & K. T. Liou (Eds.), *Disaster and development* (pp. 1–15). New York: Springer.

Kapucu, N., & Özerdem, A. (2013). *Managing emergencies and crises*. Boston: Jones & Bartlett Publishers.

Kapucu, N., Hawkins, C., & Rivera, F. (2013a). *Disaster resilience: Interdisciplinary perspectives*. New York: Routledge.

Kapucu, N., Hawkins, C. V., & Rivera, F. I. (2013b). Disaster preparedness and resilience for rural communities. *Risk, Hazards & Crisis in Public Policy, 4*(4), 215–233.

Kusenbach, M., & Christmann, G. (2013). Understanding hurricane vulnerability. In N. Kapucu & C. Hawkins (Eds.), *Disaster resiliency: Interdisciplinary perspectives* (p. 61). New York: Routledge.

Lindell, M. K. (2013). Disaster studies. *Current Sociology, 61*(5–6), 797–825.

Mittler, E. (1997). *A case study of Florida's emergency management since Hurricane Andrew*. Boulder: Natural Hazards Research and Applications Information Center, Institute of Behavioral Science, University of Colorado.

Myers, C. A., Slack, T., & Singelmann, J. (2008). Social vulnerability and migration in the wake of disaster: The case of Hurricanes Katrina and Rita. *Population and Environment, 29*(6), 271–291.

Parkinson, R. W., & McCue, T. (2011). Assessing municipal vulnerability to predicted sea level rise: City of Satellite Beach, Florida. *Climatic Change, 107*(1–2), 203–223.

Prelog, A. J., & Miller, L. (2013). Perceptions of disaster risk and vulnerability in rural Texas. *Journal of Rural Social Sciences, 28*(3), 1–31.

Stripling, M. (2013). Planning and design tool. *Disaster Planning Handbook*. Retrieved from http://www.nyc.gov/html/doh/html/em/managing-chaos.shtml

Tate, E. (2012). Social vulnerability indices: A comparative assessment using uncertainty and sensitivity analysis. *Natural Hazards, 63*(2), 325–347.

Tierney, K. (2012). Disaster governance: Social, political, and economic dimensions. *Annual Review of Environment and Resources, 37*, 341–363.

Waugh, W. L., Jr. (1994). Regionalizing emergency management: Counties as state and local government. *Public Administration Review, 54*, 253–258.

Waugh, W. L., Jr., & Liu, C. Y. (2014). Disasters, the whole community, and development as capacity building. In N. Kapucu & K. T. Liou (Eds.), *Disaster and development* (pp. 167–179). New York: Springer.

Wisner, B. (2009). *Shrink & swim: Exploring the link between capital (Social, human, institutional, natural), disaster, and disaster risk reduction*. Retrieved https://www.gfdrr.org/sites/gfdrr/files/Wisner_Long%20term%20impact%20of%20disaster%20on%20social%20capital.pdf

Zahran, S., Brody, S. D., Peacock, W. G., Vedlitz, A., & Grover, H. (2008). Social vulnerability and the natural and built environment: A model of flood casualties in Texas. *Disasters, 32*(4), 537–560.

Chapter 6
Resilience

Abstract This chapter extends the previous discussion of vulnerability (Chap. 4) by analyzing the conceptualization and elements of resilience identified in the research literature. In addition, we review federal, state, and local Emergency Management Plans with regards to resilience. We provide a detailed analysis of corresponding conceptualizations of resilience from emergency managers' perceptions including: bouncing back, restoring, avoidance, and others. We pay particular attention to differences between urban and rural settings. Increasing a community's resilience to disasters is a complex issue hinging on cooperation among multiple levels and sectors. Moreover, disaster policies and legislation must undergo a transformative process to become more proactive and predictive in its focus while understanding the importance of context, particularly differences between urban verses rural communities.

Keywords Resilience • Community disaster resilience • Urban-rural communities • Plans • Perceptions • Recovery • Measurement tools • Florida

The concept of resilience has received increasing attention in the study of disasters, particularly after several catastrophic events such as Hurricane Katrina in 2005 and the 2011 earthquake and tsunami in Japan. Indeed, there is an ongoing effort to study the ways communities are able to "bounce back" from a natural or man-made disaster (Kapucu et al. 2013). At the federal emergency management level, within the U.S., it has become an important goal to establish disaster resilient communities. Reports from the National Academy of Sciences (2012), the United Nations International Strategy for Disaster Risk Reduction (2012), and other studies (Combaz 2014) have explored, in detail, how to promote disaster resilience. Overall, these reports acknowledge the importance in understanding the interconnectedness of community's assets and invest in its future protection (National Academy of Science [NAS] 2012).

To gain support for investment, planners and officials must make sure improvements are cost effective. Monetary aid is a way to bolster the ability for administrators to provide for the needs of affected communities (Stromberg 2007). Donations are an integral part of building a resilient community and can be linked a selfish desire to feel good. The Office for Foreign Disaster Assistance found every dollar

© Springer International Publishing Switzerland 2015   69
F.I. Rivera, N. Kapucu, *Disaster Vulnerability, Hazards and Resilience*,
Environmental Hazards, DOI 10.1007/978-3-319-16453-3_6

spent on disaster relief meant four dollars saved in the long run (Stromberg 2007). Some improvements can be made after a disaster like creating multiple uses for a building. In Louisiana all the schools were fortified to transition into shelters in the event of a hazard or disaster for the community. Investment in the safety of your community does not need to happen over time, a smart approach is to slowly make changes to help planners create and enact certain measures each year until completion. Talking to families and sharing inexpensive and handy ways to make homes safer also helps the community stay safe (NAS 2012).

Knowing what a community's vulnerabilities are is crucial to increase resiliency to natural and technological hazards. Any person in a region can be exposed to hazards and knowledge of those who can be at greater risk helps save lives and mitigates damages and fatalities (Kapucu and Özerdem 2013). Those considered most vulnerable are impoverished, elderly, children, sick, or deemed to have any other disadvantage. Educating people on the predicted hazards in a certain region becomes critical to aid potential victims and give them the opportunity to protect their selves and their assets.

6.1 What Is Resilience?

On the outset of the book, we define resilience as "the ability to adapt through the redevelopment of the community in ways that reflect the community's values, and goals, and its evolving understanding of external forces with which it must contend" (Kapucu et al. 2013, p. 220). This definition builds from the National Research Council (NRC) (2009) statement of resilience as "a response to stress and can be considered as a theory that guides the understanding of stress response dynamics; a set of adaptive capacities that call attention to the resources that promote successful adaptation in the face of adversity; and strategy for disaster readiness against unpredictable and difficult to prepare for dangers" (National Research Council 2009, p. 23).

The term resilience is not new and several definitions have been put forward including Holling's (1973) view of resilience as "the measure of the ability of a system to absorb change in the face of extreme perturbation and yet continue to persist" (Brody and Gunn 2013, p. 161) to recent views suggesting resilience is governed by the interplay and dependence of ecological and social-political systems (Folke et al. 2005). Furthermore, there is an understanding of people and their communities being considered resilient when they have access to the resources needed to prevent or respond effectively to a crisis situation. For instance, the impact of Hurricane Andrew caused Florida's government to initiate structural changes to their emergency management system (Wamsley and Shroeder 1996). State and local officials had to adapt their approach to the situation. Andrew was an unpredicted disaster and is considered one of the costliest storms in United States history causing Florida a $30 billion economic impact with 52 lost lives (Hawkins and Knox 2014). "Through understanding and using continuity, coordination, and cooperation during a disaster, governmental entities create an effective, smooth operating governmental response to disaster" (Neal and Phillips 1995, p. 331).

Recently, NAS (2012) discussed the definition and origins of resiliency, in more detail: "Although resilience with respect to hazards and disasters has been part of the research literature for decades (White and Haas 1975; Mileti 1999), the term first gained currency among national governments in 2005 with the adoption of the Hyogo Framework for Action by 168 members of the United Nations to ensure reducing risks to and building resilience to disasters became a priority for governments and local communities (UNISDR 2007). The literature has since grown with new definitions of resilience and the entities or systems to which resilience refers (e.g., ecological systems, infrastructure, individuals, economic systems, communities) (Bruneau et al. 2003; Flynn 2007; Gunderson 2009; Plodinec 2009; Rose 2009; Cutter et al. 2010)" (p. 18). When analyzing the definition of resilience, the conditions of the disaster affecting individuals and communities need to be taken into account (Kapucu 2012a). Due to the importance of context, there is a need to be adaptive towards cultural differences between local officials, planners, emergency managers and other involved parties, to increase understanding and empower citizens through the inclusion of four components: social capital, community competence, information and communication, and strong economy. If a community increases their ability to coordinate response efforts, then they decrease their vulnerability, or sensitivity of their system (Kapucu 2012a).

Disaster resilience has been described as "a *process* (Norris et al. 2008; Sherrieb et al. 2010), an *outcome* (Kahan et al. 2009), or both (Cutter et al. 2008), and as a term embracing inputs from engineering and the physical, social, and economic sciences (Colten et al. 2008)" (NAS 2012, p. 18). Furthermore, it has been defined as: "the ability of countries, communities and households to manage change, by maintaining or transforming living standards in the face of shocks or stresses – such as earthquakes, drought or violent conflict – without compromising their long term prospects" (Department For Internal Development 2011, p. 6). In all, "the goal is not necessarily to bounce back to the pre-disturbance state, but rather reach a more viable or sustainable equilibrium" (Brody and Gunn 2013, p. 161). As suggested by Rivera and Settembrino (2013), resilience is not only the ability of a community or system to "bounce-back" to pre-disaster conditions, but to actually leap forward.

Determining resiliency is important for emergency managers as the field itself is focused on helping communities deal with catastrophic events. To do so, they need to incorporate aspects of technology, planning, science and management (Wilson and Oyola-Yemaiel 2001). As a profession, emergency management finds increasing responsibility placed upon the administration of the four phases of preparation, mitigation, response and recovery. FEMA has emphasized an all-hazards approach and ties managers into the prediction of negative impacts. "According to FEMA, if one looks across the range of threats we face, from fire, to hurricanes, to tornadoes, to earthquakes, to ware, one will find there are common preparedness measures that we deal with in trying to prepare for those threats" (Wilson and Oyola- Yemaiel 2001, p. 120).

6.2   Elements of Disaster Resilience

Several elements have been highlighted as important factors for disaster resilience. For instance, Combaz (2014) suggested context, disturbance, capacity to respond, and reaction are particularly important for disaster resilience. Context refers to the identification of resilience in social groups, environments, institution, and socio-political or political systems. Disturbance refers to the sudden events or "shocks" impacting the vulnerability of the system. Examples of these shocks include natural disasters such as floods, droughts or earthquakes along with other political and economic shocks like volatility or unrest resulting in fighting or violence. In addition to shocks, disturbances also include long-term stresses to the system and have the potential to increase the vulnerability of actors within the system. These stresses include: degradation of resources, climate change, economic decline, demographic changes, loss of agricultural production, and political instability.

The emphasis in addressing disturbances is finding a response to the question 'resilience to what?' Capacity to response includes exposure to risk (e.g. the magnitude and frequency of shocks), sensitivity of the system to respond to a given shock or stress, and the adaptive capacity of actors (including communities, governments, individuals, institutions, organizations, and regions) to anticipate, plan, react and learn from stresses or shocks. Similarly, the NAS (2012) outlines key elements for a national disaster resilience framework. These elements incorporate: "public awareness of and responsibility for managing local disaster risk, establishing the economic and human value of resilience to help encourage long-term commitments to enhance resilience; tools or metrics for monitoring progress toward resilience and to understand what resilience looks like for different communities; creating local, community capacity, because decisions and the ultimate resilience of our nation derive from the bottom-up community efforts; identifying sound, top-down government policies and practices to build resilience; identifying and communicating the necessary roles and responsibilities between communities and all levels of government in building resilience, including gaps in and challenges to communications and actions among these actors" (p. 20).

6.3   Disaster Resilience: Plans and Perceptions

As the theme of the book suggests it is important to consider emergency personnel views on resilience and the elements and factors discussed. Beginning with a brief conversation on the history of comprehensive laws. The first bit of legislation was passed in 2000 requiring local jurisdictions to create comprehensive plans for emergency management (Tierney 2012). Following this legislation, the Federal Response Plan (FRP) provided a foundation for administrative efforts as it guides emergency management through the incorporation of emergency support functions and an emphasis on knowledge, skills, and abilities of managers (Wilson and Oyola- Yemaiel 2001). Branching off of the FRP, comprehensive plans have surfaced to analyze disaster response and recovery through a context-based lens. More specifically, the National Disaster Recovery Framework, made available in 2011, informs the U.S. on guidelines for post-disaster recovery (Tierney 2012).

Broadening the spectrum, the U.S. is not the only country working to create CMPs for their communities. The International Federation of Red Cross and Red Crescent Societies (IFRC) generated a program focused on community resiliency with a program incorporating three key objectives: "providing a framework that guides IFRCs community resilience programming at scale; communicating, articulating and advocating IFRCs position in relation to community resilience; and identifying financing methodologies that support community resilience programming" (International Federation of Red Cross 2014, p. 1). With these objectives, IFRC hopes to be more intentional on the areas of knowledge, practice, analysis and comparisons, and recommendations on method and performance indicators.

The framework and objectives are created with the intention of increasing community resiliency within a humanitarian and developmental perspective. Cutter (2013) promotes the global perspective due to five pillars she examined: "to leave no one behind; put sustainable development at the core; transform economies for jobs and inclusive growth; build peace and transparent and accountable institutions; and forge new global partnerships. If such a transformative shift takes place, by 2030 the world would see increased resilience and improved quality of life" (p. 77). Initials steps to increasing resiliency include: "(1) build on local community knowledge of disasters; (2) prepare and deliver effective warning messages (news that will reach all diverse populations, including blind, deaf, and minority individuals); (3) develop shelters; (4) anticipate the need for translators, child care, medical equipment, and other resources; and (5) avoid wasting resources on unnecessary supplies due to bias" (Kapucu and Özerdem 2013, p. 28).

In the next section, we use Central Florida as a case example to explore which elements of resilience are already in place in the plans and perceptions of those involved in disaster emergency management.

6.4 Analysis of Florida Comprehensive Management Plan

A review of the 2012 State of Florida Comprehensive Management Plan stating the plan "employs the strategic vision of Presidential Policy Directive 8 (PPD- 8), to strengthen resiliency by involving partners at all levels of government as well as with nongovernmental organizations (NGOs) and the private sector in the planning process" (p. 4). Therefore, although not defined, the CMP invokes the elements of resilience previously reviewed and strives to be a guideline for all elements of the emergency management cycle including notification, mobilization, detection and activation. Along with response, recovery, and mitigation, CMPs achieve "state goals on two tiers of implementation: (1) state agencies implementing legislated state goals and objectives and (2) local governments implementing state agency directives" (Deyle et al. 2008, p. 350).

When it comes to emergency preparation, mitigation, response and recovery, there is an essential human element. Disasters not only affect people, but people affect them as well. The aspect of mobilization becomes imperative when disasters strike as emergency management hinges on the cooperation between federal, state and local officials along with available volunteers (Volunteer Florida 2014). In regards to emergency support functions, the Florida Division of Emergency Management established a function focused specifically on volunteer efforts. ESF-15, the volunteer element, is critical. Volunteer Florida (2014) is one organization who coordinates with Voluntary Organizations Active in Disaster (VOAD) to help disseminate crucial information, operate web-based technological efforts, provide staffing, collecting data, training, presenting and assist in fiscal management before, during and after disasters. These volunteers became an essential piece in determining resiliency for an area. "During times when disasters are coordinated between different of levels of government the ESF-based structure is specifically important. The standardization of resource grouping as well as the responsibilities of actors leads to a more streamlined response and recovery process" (Kapucu 2012b, p. s44).

At the county level, we examined the CMPs of Orange (urban) and Volusia (rural) counties and found no explicit mention of resilience. As expected their plans mirror the state plan in relation to the emergency management cycle. Not surprisingly, when asked to define resilience, the open-ended responses from the survey participants closely linked to the disaster functions detailed in their CMPs particularly in the view of resilience as preparedness, "bouncing-back," and recovery. Below are some examples of the responses given:

- Preparedness

    - The ability to have a plan in place and depending on the impact of the disaster be flexible enough to change and adapt to new procedures.
    - Doing our best to be prepared for all hazards and also to help prevent them
    - The ability for a community to get "back to normal" following a disaster. This is completed through adequate preparation and training at all levels, from the community through government and private

- Bouncing Back

    - Being able to "bounce back" from a disaster.
    - How quickly a community to "bounce back" from a disaster.
    - The ability to return to normal status in a timely manner relevant to the amount of damage sustained.

- Recovery

    - The ability to recover relatively quickly and efficiently from a disaster back to normal community operations.
    - The ability to get through and recover from a disaster in which the organization can then resume normal operations.
    - The capability of a community/organization to respond and quickly recover to pre-disaster conditions.

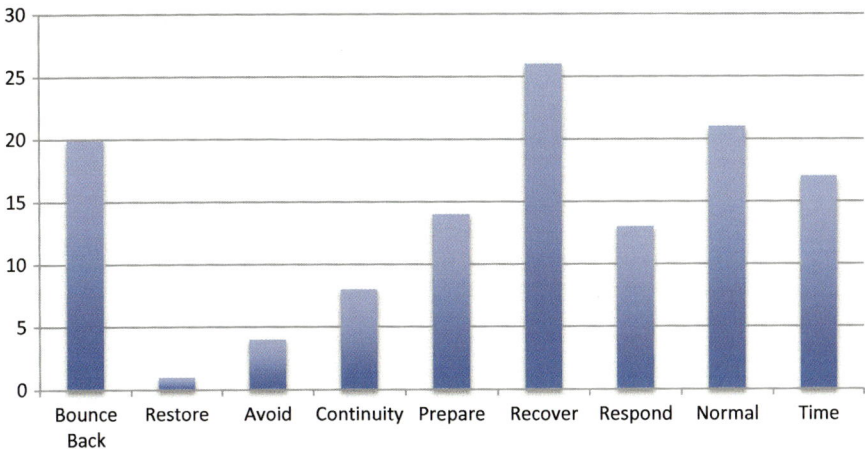

Fig. 6.1 How do you define resilience?

Illustrated in Fig. 6.1 are the open-ended responses for all the counties. In addition, the responses are compared from urban verses rural counties. In all, respondents defined disaster resilience as the ability to prepare for, respond to, and recover from a disaster. Particularly, there was an emphasis on returning to "normal" ("Bouncing Back") quickly and efficiently.

Contrasting the concept of 'normal,' it has become more apparent how bouncing back from a disaster increases in difficulty with each new disaster (Deyle and Smith 1998). Most local governments tend to focus on emergency response in a reactive fashion. Therefore, the cost of returning to 'normal' increases significantly. Conversely, the promotion of generating CMPs can help local governments lower the economic impact to their communities. With Florida being one of the higher states to mandate emergency management plans, our communities are more capable of increasing resiliency and being about to bounce back when future disasters strike (Deyle and Smith 1998). For example, Hurricane Katrina in 2005 caused the state of Louisiana to become painfully aware of gaps and issues within their emergency response plans. "Thus, effective local implementation of state goals is determined by the quality of the plan policies adopted by local governments, how local officials translate those policies into development controls and other growth management measures, and how they enforce those measures" (Deyle et al. 2008, p. 350).

Unique environmental aspects of Hurricane Katrina included the structure of the levee system and issues with coastal erosion. Seeing as Florida is a state dominated by coastal areas, it is not surprising for emergency management policies and legislation to analyze related issues in land-use and development (Tierney 2012). Predictions express Florida will increase its coastal population 25 % by 2050 (Aerts et al. 2014). Compound the increased demographic diversity with projected sea-level rise and climate change and there is a growing chance of more severe impacts

Fig. 6.2 Comparison of
urban and rural responses

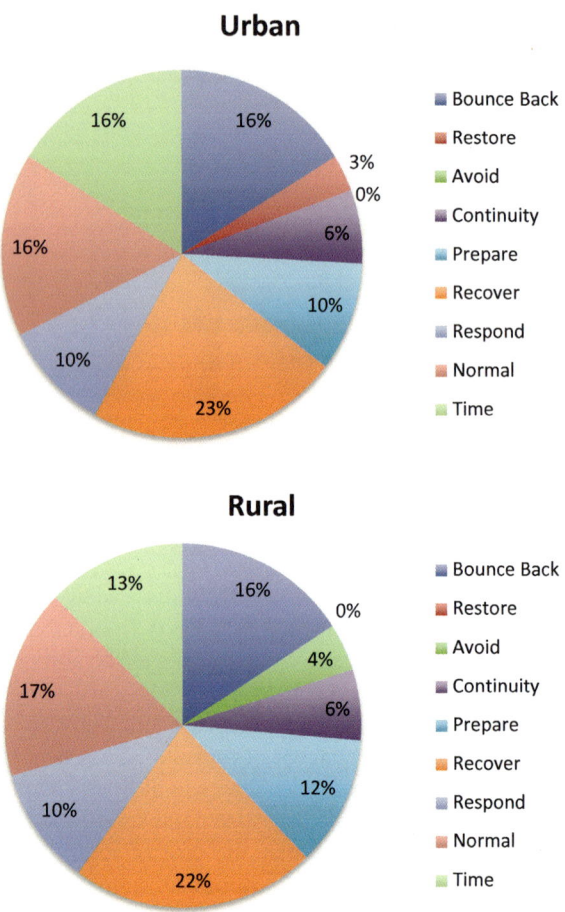

from disasters such as large-scale floods. Recovery efforts are going to intensify in
difficulty if emergency management personnel continue to view resilience as needing short-term solutions.

The responses (see Fig. 6.2) encompass elements in line with the National
Academies Committee on Increasing National Resilience definition of *resilience* as
"the ability to prepare and plan for, absorb, recover from, and more successfully
adapt to adverse events" (NAS 2012, p. 16). For respondents in urban counties prepare constituted 12 % of the responses, recover 23 %, and bounce back 16 %.
Similarly, respondents in rural counties prepare constituted 12 % of the responses,
recover 22 %, and bounce back 16 %.

With recovery being the higher statistic, it is logical to focus on comprehensive
plans to reduce the negative impacts of disasters. Building the capability for a community is done through "pre-event activities, such as hazards and vulnerability
assessments; land-use regulations; building code development, adoption, and

enforcement; warning systems; and education and training programs" (Tierney 2012, p. 344). Effective administration of disaster phases helps constituents to better understand the situation and to interpret relevant policies and procedures. Some post-disaster measures to further support planning efforts can include "short- and longer-term recovery programs, as well as the formulation and implementation during disaster recovery of interventions designed to reduce future disaster losses and promote sustainability" (Tierney 2012, p. 344). For Florida, comprehensive management began in 1990 and furthered as a response to Hurricane Andrew, but there were issues during implementation because the document was not as detailed as it needed to be (Mittler 1997).

Hurricane Katrina can be used to exemplify the need for focusing on more long-term resiliency efforts. Many disaster policies focused on the band-aid effect, or short-term needs, without keeping in mind the potential for economic deficits to occur if the issues were not processed through a lens of sustainability (Cutter 2013). It is up to emergency management administration to adjust the perspective of their disaster response and recovery policies (See Fig. 6.3).

The crucial component to these post- and pre-disaster activities is to make sure context is at the forefront. Social disparities can become a detriment if not acknowledged. Potential assessment tools include generic indexes like the Genuine Progress Indicator, World Development Indicators, Human Development Index, Environmental Vulnerability Index, Environmental Sustainability Index, Local Disaster Index, and the Disaster Deficit Index (Tierney 2012).

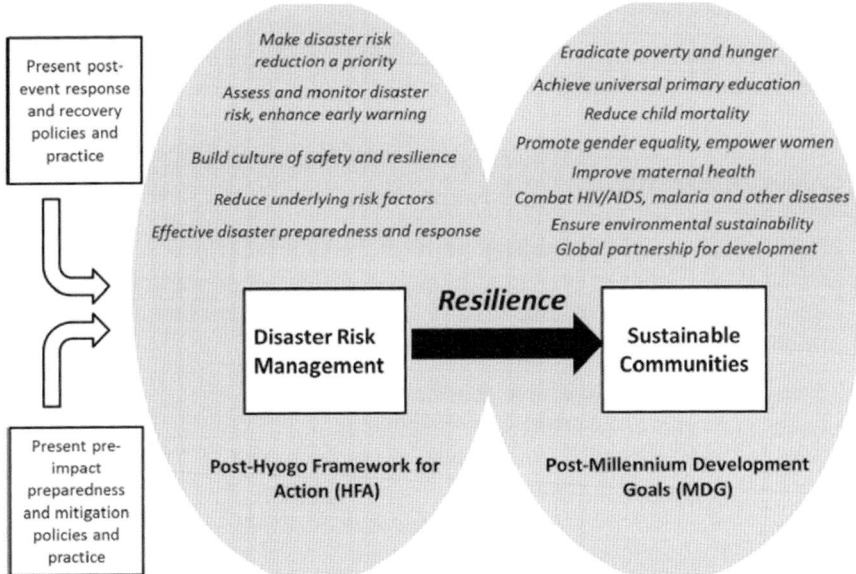

Fig. 6.3 The path for a more resilient community (Cutter 2013)

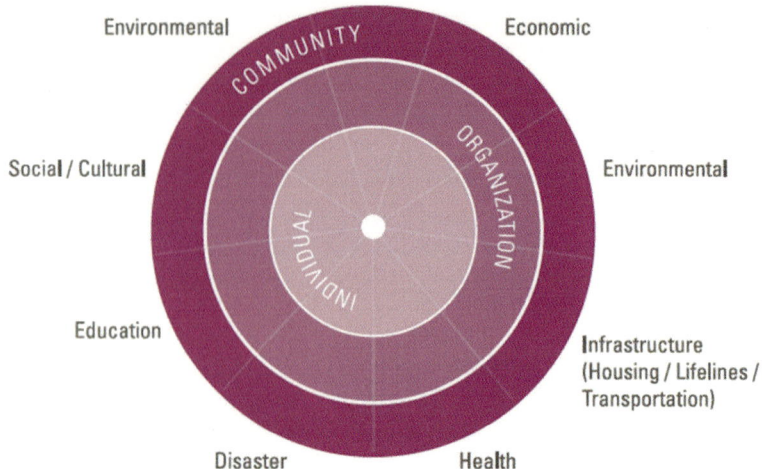

Fig. 6.4 Components of local government self-assessment tool (UNISDR 2012)

Cutter and Emrich (2006) propose utilizing the Social Vulnerability Index (SVI) as one of the tools used to examine an area for the intersection of social vulnerabilities to the environmental hazards and produce a quantitative result. These measurements provide a way to analyze issues of disaster vulnerability along with governance issues. In addition to the SVI, UNISDR proposes a Local Government Self-Assessment Tool (see Fig. 6.4 for tool components) to:

- Help local governments engage with different stakeholders to map and understand existing gaps and challenges in disaster risk reduction in their city or locality.
- Set a baseline and develop status reports for cities and municipalities that have committed to the Making Cities
- Resilient Campaign and its Ten Essentials.
- Complement information gathered through the national Hyogo Framework for Action (HFA) monitoring system by providing local-level information. Cities can chose to share their results with national HFA focal points as part of the national reporting process (p. 78).

Concentrating on rural communities, there are several factors increasing the vulnerability of the area and causing issues in building resilience as rural communities are seen as those susceptible to disasters and hazards with an assumption of an inability to recover as quickly as more urban areas (Durant 2011). Several marginalization factors include limited economic, political, social and human resources (Morrow 1999). Moreover, there is a geographic issue as well due to the distance between rural and urban communities (Kapucu et al. 2013). Although there are indications of economic pitfalls for rural communities, there is the recognition of community members coming together when disasters occur. In fact, Tobin (1999) finds these areas to have a greater chance at gaining resiliency within the following characteristics:

- Lowered levels of risk to all members through reduced exposure to the geophysical event; reduced levels of vulnerability for all members of society;
- Ongoing planning for sustainability and resilience;
- High level of support from responsible agencies and political leaders;
- Incorporation of partnership and cooperation at different governmental levels;
- Strengthened networks for independent and interdependent segments of society; and planning at the appropriate level (p. 17).

It seems some of the disaster resilience elements previously discussed were present or acknowledged by the survey participants (particularly preparedness, bouncing back, and recovery). Other elements of resilience, particularly issues relating to apathy and complacency, communication issues, and funding will be discussed in the next chapter.

6.4.1 Conclusion

Increasing a community's resilience to disasters is a complex issue hinging on cooperation among multiple levels and sectors. Government and constituents must coordinate efforts for preparation, mitigation, response and recovery efforts intertwining risk management and sustainable development. Moreover, disaster policies and legislation must undergo a transformative process to become more proactive and predictive in its focus while understanding the importance of context. Each community differs in its make-up and may require specific programs. For example, there are differences in vulnerabilities of urban communities versus rural including limited economic, political, social and human resources. Tools to measure a community's social context and vulnerabilities are an important aspect of governance, as resilience to disasters must be enhanced to ensure the livelihoods and prosperity of future generations.

References

Aerts, J. C., Botzen, W. W., Emanuel, K., Lin, N., de Moel, H., & Michel-Kerjan, E. O. (2014). Evaluating flood resilience strategies for coastal megacities. *Science (New York, NY), 344*(6183), 473–475.

Brody, S. D., & Gunn, J. R. (2013). Chapter 9: Examining environmental factors contributing to community resilience along the Gulf of Mexico Coast. In N. Kapucu, C. Hawkins, & F. Rivera (Eds.), *Disaster resiliency: Interdisciplinary perspectives* (pp. 160–177). New York: Routledge. https://www.gov.uk/government/uploads/system/uploads/attachment_data/file/186874/defining-disaster-resilience-approach-paper.pdf

Bruneau, M., Chang, S. E., Eguchi, R. T., Lee, G. C., O'Rourke, T. D., Reinhorn, A. M., Shinozuka, M., Tierney, K., Wallace, W. W., & von Winterfeldt, D. (2003). A framework to quantitatively assess and enhance the seismic resilience of communities. *Earthquake Spectra, 19*(4), 733–752.

Colten, C. E., Kates, R. W., & Laska, S. B. (2008). Three years after Katrina: Lessons for community resilience. *Environment, 50*(5), 36–47.

Combaz, E. (2014). *Disaster resilience: Topic guide.* Birmingham: GSDRC, University of Birmingham.

Cutter, S. L. (2013). Building disaster resilience: Steps toward sustainability. *Challenges in Sustainability, 1*(2), 72–79.

Cutter, S. L., & Emrich, C. T. (2006). Moral hazard, social catastrophe: The changing face of vulnerability along the hurricane coasts. *The Annals of the American Academy of Political and Social Science, 604*(1), 102–112.

Cutter, S. L., Barnes, L., Berry, M., Burton, C., Evans, E., Tate, E., & Webb, J. (2008). A place-based model for understanding community resilience to natural disasters. *Global Environmental Change, 18*, 598–606.

Cutter, S. L., Burton, C. G., & Emrich, C. T. (2010). Disaster resilience indicators for benchmarking baseline conditions. *Journal of Homeland Security and Emergency Management, 7*(1), Article 51.

Department For Internal Development. (2011). *Defining disaster resilience: A DFID approach paper.* DFID. https://www.gov.uk/government/uploads/system/uploads/attachment_data/file/186874/defining-disaster-resilience-approach-paper.pdf

Deyle, R. E., & Smith, R. A. (1998). Local government compliance with state planning mandates: The effects of state implementation in Florida. *Journal of the American Planning Association, 64*(4), 457–469.

Deyle, R. E., Chapin, T. S., & Baker, E. J. (2008). The proof of the planning is in the platting: An evaluation of Florida's hurricane exposure mitigation planning mandate. *Journal of the American Planning Association, 74*(3), 349–370.

Durant, T. J. (2011). The utility of vulnerability and social capital theories in studying the impact of Hurricane Katrina on the elderly. *Journal of Family Issues, 32*(10), 1285–1302.

Flynn, S. (2007). *The edge of disaster: Rebuilding a resilient nation.* New York: Random House.

Folke, C., Hahn, T., Olsson, P., & Norberg, J. (2005). Adaptive governance of social–ecological systems. *Annual Review of Environment and Resources, 30*, 441–473.

Gunderson, L. (2009). *Comparing ecological and human community resilience* (Research Report 5). Oak Ridge: Community and Regional Resilience Initiative. Available at http://resilientus.org/library/Final_Gunderson_1-12-09_1231774754.pdf

Hawkins, C., & Knox, C. C. (2014). Disaster events and policy change in Florida. In N. Kapucu & K. T. Liou (Eds.), *Disaster & development: Examining global issues and cases.* New York: Springer.

Holling, C. S. (1973). Resilience and stability of ecological systems. *Annual Review of Ecology and Systematics, 4*, 1–23.

International Federation of Red Cross and Red Crescent Societies. (2014). *Terms of reference: Community resilience performance measurement methodology and standard indicators.* Retrieved from http://www.ifrc.org/PageFiles/99169/TOR_Final_23June.pdf?epslanguage=en

Kahan, J. H., Allen, A. C., & George, J. K. (2009). An operational framework for resilience. *Journal of Homeland Security and Emergency Management, 6*(1), Article 83. Available at: http://www.bepress.com/jhsem/vol6/iss1/83/

Kapucu, N. (2012a). Disaster and emergency management systems in urban areas. *Cities: The International Journal of Urban Policy and Planning, 29*(s1), 41–49.

Kapucu, N. (2012b). Disaster resilience and adaptive capacity in Central Florida, US, and in Eastern Marmara Region, Turkey. *Journal of Comparative Policy Analysis: Research & Practice, 14*(3), 202–216.

Kapucu, N., & Özerdem, A. (2013). *Managing emergencies and crises.* Boston: Jones & Bartlett Publishers.

Kapucu, N., Hawkins, C. V., & Rivera, F. I. (Eds.). (2013). *Disaster resiliency: Interdisciplinary perspectives*. New York: Routledge.

Mileti, D. S. (1999). *Disasters by design: A reassessment of natural hazards in the United States*. Washington, DC: Joseph Henry Press.

Mittler, E. (1997). *A case study of Florida's emergency management since Hurricane Andrew*. Boulder: Natural Hazards Research and Applications Information Center, Institute of Behavioral Science, University of Colorado.

Morrow, B. H. (1999). Identifying and mapping community vulnerability. *Disasters, 23*(1), 1–18.

National Academy of Sciences (NAS). (2012). *Disaster resilience: A national imperative*. Washington, DC: The National Academies Press.

National Research Council (NRC). (2009). *Applications of social network analysis for building community disaster resilience*. Washington, DC: The National Academies Press.

Neal, D. M., & Phillips, B. D. (1995). Effective emergency management: Reconsidering the bureaucratic approach. *Disasters, 19*(4), 327–337.

Norris, F. H., Stevens, S. P., Pfefferbaum, B., Wyche, K. F., & Pfefferbaum, R. L. (2008). Community resilience as a metaphor, theory, set of capacities, and strategy for disaster readiness. *American Journal of Community Psychology, 41*, 127–150.

Plodinec, M. J. (2009). *Definitions of resilience: An analysis*. Oak Ridge: Community and Regional Resilience Initiative. Available at http://www.resilientus.org/library/CARRI_Definitions_Dec_2009_1262802355.pdf

Rivera, F. I., & Settembrino, M. R. (2013). Chapter 3: Sociological insights on the role of social capital in disaster resilience. In N. Kapucu, C. Hawkins, & F. Rivera (Eds.), *Disaster resiliency: Interdisciplinary perspectives* (pp. 48–60). New York: Routledge.

Rose, A. (2009). *Economic resilience to disasters* (CARRI Research Report 8). Oak Ridge: Community and Regional Resilience Initiative. Available at http://www.resilientus.org/library/Research_Report_8_Rose_1258138606.pdf

Sherrieb, K., Norris, F. H., & Galea, S. (2010). Measuring capacities for community resilience. *Social Indicators Research, 99*(2), 227–247.

Strömberg, D. (2007). Natural disasters, economic development, and humanitarian aid. *The Journal of Economic Perspectives, 21*, 199–222.

Tierney, K. (2012). Disaster governance: Social, political, and economic dimensions. *Annual Review of Environment and Resources, 37*, 341–363.

Tobin, G. A. (1999). Sustainability and community resilience: The holy grail of hazards planning? *Global Environmental Change Part B: Environmental Hazards, 1*(1), 13–25.

United Nations International Strategy for Disaster Risk Reduction (UNISDR). (2012). *How to make cities more resilient: A handbook for local government leaders*. Geneva: United Nations International Strategy for Disaster Risk Reduction (UNISDR).

United Nations International Strategy for Disaster Reduction Secretariat (UNISDR). (2007). *Hyogo framework for action 2005–2015: Building the resilience of nations and communities to disasters*. Geneva: UN/ISDR. Available at http://www.unisdr.org/files/1037_hyogoframeworkforactionenglish.pdf

Volunteer Florida. (2014). *Emergency management: About ESF_15*. Retrieved from http://www.volunteerflorida.org/emergency-management.com

Wamsley, G. L., & Shroeder, A. D. (1996). Escalating in a quagmire: The changing dynamics of emergency management policy subsystem. *Public Administration Review, 56*(3), 235–244.

White, G. F., & Haas, J. E. (1975). *Assessment of research on natural hazards*. Cambridge, MA: MIT Press.

Wilson, J., & Oyola-Yemaiel, A. (2001). The evolution of emergency management and the advancement towards a profession in the United States and Florida. *Safety Science, 39*(1), 117–131.

Chapter 7
The Path to Resilience

Abstract In this chapter, we discuss some of the challenges to resilience utilizing a conceptual framework (Adaptive Resilience & Community Capital Framework) that includes several elements essential to disaster resilience. We build on the Adaptive Resilience & Community Capital framework by providing a detailed analysis of the emergency managers and other non-profit and community group responses with regards to what they perceived to be obstacles to resilience, particularly issues relating to apathy and complacency, communication issues, and funding, among others. In all, increasing a community's resilience capability consists of a variety of aspects such as socioeconomic, complacency, shifting demographics and resources in terms of shelters and facilities. To overcome these barriers, attention must be given to the nuances of a community, including communication issues, seasonal residents, mistrust of the government, and its collective capital. Furthermore, communication and coordination among agencies is a critical aspect to disaster resilience.

In the last chapter we discussed context, disturbances, capacity, and sensitivity as key elements for disaster resilience. In addition, we identified and discussed preparedness, bouncing back, and recovery as key factors in the definition of resilience among emergency management personnel. Yet, as witnessed in our discussion of all the factors identified in both national (National Academy of Sciences (NAS), Disaster resilience: a national imperative. The National Academies Press, Washington, DC, 2012) and international (Combaz E, Disaster resilience: topic guide. GSDRC, University of Birmingham, Birmingham, 2014) frameworks, there are several outstanding issues considered as challenges or roadblocks to the path of resilience (see Fig. 6.1, a conceptual map including several elements essential to disaster resilience for both rural and urban communities).

Keywords Resilience • Community capital • Adaptive resilience • Funding • Brain loss • Social factors • Mistrust • Communication • Florida

7.1   Challenges to Resilience

In this section, we concentrate our discussion on the elements perceived as potential challenges to resilience. A key element of the framework is integrating learning and adaption into the phases of disaster management and all hazard perspective in response to stressors. In this context, disaster resilience is defined as the ability to adapt through the redevelopment of the community in ways which reflect the community's assets, values and goals, and its evolving understanding of external forces deemed prior struggles (Kapucu et al. 2013). This framework also incorporates the evolving understanding of how to move in a desirable direction to sustain positive community structures and functions and remove the negative ones after a disaster. This refocuses attention to the "community capital," or community assets, to be utilized in planning for and implementing activities during the recovery process. This capital is partly influenced by local socio-economic conditions along with external forces such as state- and federal-level policies and resources, level of social capital, and the availability of civil society organizations.

Examining Fig. 7.1, there are several components incorporated into the process of responding to specific stressors (i.e. degradation of resources, climate change, economic decline, demographic changes, loss of agricultural production, and political instability). With a focus on the adaptive capacity of a community, some

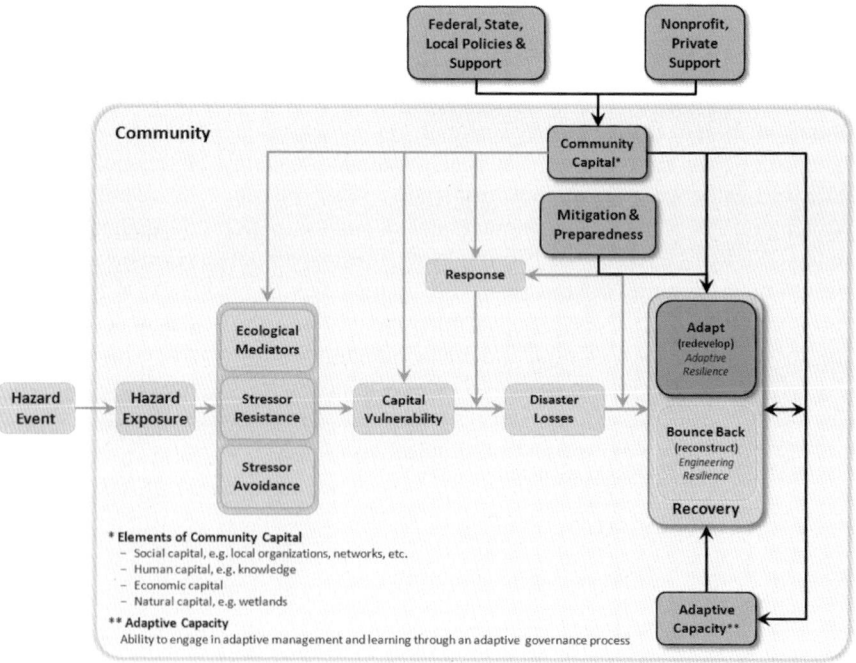

Fig. 7.1 Adaptive resilience & community capital (Adapted from Kapucu et al. 2013)

begin with the policies and procedures currently existing and were generated by federal, state, and local officials. The guidelines combine with the support from the nonprofit and private sector into the community capital. These resources can help emergency managers predict the ability for their area to return to a state of normal once a hazard or disaster affects them. Although you cannot return to the time before the situation, there is a question of how much the community can be moving forward once more (Rivera and Settembrino 2013).

Proactive approaches to increasing resiliency begin with understanding how the area is susceptible to triggers like hazards (i.e. floods, industrial accidents, etc.) and vulnerabilities (i.e. environmental, social, physical and economic conditions) (Henstra 2010). The discovery of how unique each community is through examining aspects like socioeconomics, environmental nuances, and predictability of disaster situations assists with building capacity. Adaptive capacity is a way to analyze exposure to risk (e.g. the magnitude and frequency of shocks), sensitivity of the system to respond to a given shock or stress, and the ability of involved agencies (including communities, governments, individuals, institutions, organizations, and regions) to anticipate, plan, react o and learn from stresses or shocks (Combaz 2014).

From this perspective, disaster resilience is considered a function of the community's adaptive capacity and helps the community engage in adaptive management and continuous learning as recovering from each event becomes more and more difficult (Deyle and Smith 1998). Adaptive learning can enhance community capital and develops local capacity in responding and recovering from disasters. This in turn influences disaster resilience through mitigation and preparedness enabling a more effective response to and recovery from disasters. We also expect lessons learned from response and recovery experiences will help with the design of more resilient and sustainable communities in the future. To minimize disaster losses, it is important to understand the vulnerability of capital and its ability to prepare and implement both response and recovery activities. The reciprocal relationship between adaptive capacity and community capital feeds back to disaster resilience through mitigation, preparedness, response, and recovery. All disasters are local. The rising expectations of disaster resilience place burdens on local government, especially in rural communities, responsible for managing disasters. Emergency managers must learn to partner with other organizations, community members, and jurisdictions. In short, risks are shared and a variety of community assets should be leveraged under stressful environments caused by disasters. Disasters are also complex in nature requiring interdisciplinary perspectives from different sectors in communities. To achieve a sense of resiliency, Tobin (1999) projects the following:

- Lowered levels of risk to all members through reduced exposure to the geophysical event; reduced levels of vulnerability for all members of society;
- Ongoing planning for sustainability and resilience;
- High level of support from responsible agencies and political leaders;
- Incorporation of partnership and cooperation at different governmental levels;
- Strengthened networks for independent and interdependent segments of society; and planning at the appropriate level (p. 17).

7.2   Perceptions of Obstacles to Disaster Resilience

There were several responses in our survey instrument with regards to the respondent's perceptions of obstacles to disaster resilience. Open-ended responses emphasize funding, complacency, and apathy from the community. Individual responses are listed below:

> "The general public does not plan sufficiently," "Funding and complacency," "Cost," and "Time and money." "Time to get people involved and make them aware of the resources and money to make it happen," "Reductions in resources," "Aging infrastructure," "Money, convincing people of the need to spend more money to make it resilient when they can spend just enough money to get by."

The following response captured all the identified perceived obstacles to resilience:

> A lack of awareness on the part of its citizens on the role that their government plays in disaster response. Antipathy towards local, state, and the federal government has grown to such an extent that it has become accepted wisdom that government is a nuisance. While most citizens would acknowledge the importance of the role that first responders play during a disaster for example, they also fail to make the connection that sufficient tax revenues, and a functioning government, are required to coordinate an effective disaster response.

Other issues identified included the perception of lack of cooperation and training with certain agencies. For instance, a participant stated "lack of co-operative planning and training" and another suggested "Creating a system where all agencies work together." These concerns and others were also expressed in our focus groups data, which we discuss in the next section.

7.3   Perceived Challenges to Disaster Resilience

7.3.1   Principal Themes

We analyzed the focus groups data and found several themes to the perceived challenges to disaster resilience in their respective organizations and communities. The most common issues expressed were: funding and the economy, complacency/apathy from the community, brain loss and people leaving, and issues with adequate facilities and shelters. We expand on these themes below.

At the time the interviews took place (from November 2011 to March 2012), the U.S. economy was rippling from the effects of the 2008 economic meltdown. As a consequence, federal, state, and local governments slashed budgets and the effects were felt by emergency management agencies. The struggling economy, alongside the relative calm hurricane seasons in Florida since 2004, made funding a primary concerning for most agencies. For instance, participants from Brevard, Lake, Orange, and Sumter counties named the economy and lack of funding as a particular challenge for disaster resilience. As participants from Brevard County stated:

"Every county is struggling to find dollars to support emergency response, whether its law, fire, emergency medical, emergency management, public works, health departments, volunteers."

> It's been quite a while since we had a hurricane come through here, and you can see the interest in preparedness and mitigation and things like that drops off the further away you get from a significant event. So that's, uh, and right now with the budget challenges and everything that everybody is facing because of the economy, and it's been five or six, or eight years (X: eight years), eight years since we had a hurricane with a major impact in the county. It's, it's a big challenge.

In Lake County, participants express the same concerns: "In terms of, um, disaster resiliency, and especially dealing with uh, with Florida communities that have been hit so hard in the economic crisis, we've cut our resources so much until some of the basic resources that provide the infrastructure that would step in to provide that kind of resiliency is no longer accessible." Similarly in Orange County participants stated: "Yeah, I think a lot of that, um, is somewhat attributed to the um, the overall atmosphere of the economy, because what happens when people are under stress from a financial standpoint, they tend to retract unto themselves and, they go into more of a survival mode, self survival mode." Finally, a participant in Sumter County shared: "And because of the funding, and because of other things, um, or lack their of, this, um, being prepared for an emergency is not a priority for most people."

Unfortunately, public complacency is much like the boy who cried wolf in Aesop's Fables (O'sullivan 2003). The story revolves around a young sheep watcher whose responsibility was to warn villagers when a wolf was in the area. The boy becomes bored at one point and begins tricking the locals into believing there is a threat when there is none. The villagers soon fail to take the boy's antics seriously and a wolf does attack the herd. This fable has been used as a cautionary tale of lying; however, it relates to the public's habit of not taking disaster warnings seriously. Each year, for example, Floridians are warned about hurricane season and are urged to prepare themselves. Yet, there seems to be a growing number of individuals who refuse to properly prepare. When the disaster does strike, then those individuals can become more negatively impacted.

Funding challenges were followed by a perception of apathy and complacency from the community with regards to preparedness and disaster resilience. Statements from participants in Brevard, Orange, and Sumter counties attest to this sentiment:

> Well right now I think its complacency because we haven't had an event since the tornadoes in 2004. We have a lot of people that are like "Okay, we did that so, that ain't gonna happen again for twenty more years so why bother to get ready and get prepared." I mean, I think that if were we going to have an event, if we had events this last year, I think we would have been in the same situation we were in back in 2004. Many people would not have been prepared, because they just get into this complacency that, you know, "it's not gonna happen again" and that was quirky thing, and so, it's gonna make it harder for us to bounce back, because uh, people, are not prepared. They don't even have the water in place and things to take care of their family for three days (Orange County).

"Complacency. And then reliability, or relying on government to do everything for you. But complacency; people get comfortable in their ways. We haven't had a hurricane since the 05 season, or tropical storm Faye in 08. So people get complacent, they get comfortable, you know, they have this attitude that it's not gonna happen to me" (Seminole).

Wang and Kapucu (2008) found a community's ability to ignore threat warnings leads to parallel between communication and coordination. If a community fails to take warnings seriously, then they are at a higher risk of being vulnerable during a disaster situation. Although, there has not been a clear definition of public complacency and relevant theories, the issue is prominent for emergency managers as their role revolves around keeping their communities aware of risks and vulnerabilities.

Moreover, complacency can be linked to the idea of responsibility. Citizens rely on their local officials and emergency managers to be experts. Lack of coordination can be seen when the local community sits and waits for someone to guide them verses coming forward due to a discomfort with the situation. "Regardless of the type of warning system that is chosen, its effectiveness partly depends on the people's willingness to take action…people are more likely to take action if they have been previously educated" (Henstra 2010, p. 240).

Another theme we discovered was the issue of brain loss or people leaving their respective communities mainly due to economic issues. For instance, Brevard County experienced major job losses with the restructuring of the federal space program, while other counties have experienced loss of experienced emergency personnel who have yet to be replaced due to the dire economy and lack of funding, such as the case in Lake County. Participants shared the following as challenges to disaster resilience: "…Especially recently, because of the changes in the space program we've had a tremendous change over in the county. A lot of people are leaving" (Brevard County).

> I think she brings up a really good point because the last major impacts we had were around 2004 and 2005, and there's also been a brain loss of people that have retired, who we've lost their battle experience and we haven't gotten the younger group in because of the economy. We don't have the resources to train (them) you know in damage assessment. And you know that's the biggest fear, if we have an event now, in the low economic conditions, we don't have the resources to do damage assessment to get the wheels rolling. I'm afraid we'll be stuck days into the event, just waiting to find out if it's a major disaster, just how much impact we've received to get to the different levels of that (emergency assistance/funding) because we don't have the 10 damage assessors, we may have one or two and he or she is going to be really tapped out to be able to get that (Lake County).

The last major theme was with adequate facilities and shelters. These were concerns by participants in Lake and Sumter Counties. For example, one participant expressed: "I think specifically for our community, one of the challenges is the facility that we operate in for emergency operations. Uh, you know that's a crucial issue because, you know, the board is going to make a decision, hopefully make a decision on February 28th, but their decision relates to the public whether there is an importance to being prepared or stuff like that. There are 60 counties out of 67 that have specially designed EOC's. We're not one of them" (Lake County). Similarly, a focus group participant in Sumter County shared the following when asked about challenges to resilience: "I think it would be hardened facilities. If there is an event, like a hurricane specifically, the hardened facilities in the south end of the county fill up pretty quick; especially the ones for the special needs."

7.3.2 Other Challenges to Resilience Themes

In addition to the principal themes already discussed, we found several topics mentioned individually in the county focus groups. Below we discuss the discovered ideas in each individual County focus group.

7.3.3 Brevard

In the Brevard County focus groups, participants discuss the following issues as challenges to resilience: lack of volunteers and issues with transient communities, in particular snowbirds. One participant shared the concern with lack of volunteers as a challenge to resilience: "Well, I'll speak for my organization. We have, uh, a limited number of volunteers in service, to help families who are affected by disasters, fires mostly. And we are stretched too thin to cover the extent, or the length of this county." Another focus group participant expressed concern with some the issues regarding transient residents: "Part of that (apathy) is due to the transient nature of Floridians. You know? How many people live here now but didn't live here when those hurricanes came through and affected us? There's a lot of movement in Florida, more so than other states. And snow birds as well." The movement was perceived as an obstacle for effective preparedness and sufficient volunteers, as stated in the following quote:

> Especially in the summer with snow birds flying north, yeah, we, we have fewer volunteers available to us in the summer months. They go up north for the summer months. So yeah, that's another challenge. Um, you know, even some of our volunteers, they're with us in the fall and winter and spring, and they're gone in the summer." Another respondent added: "So they might be here for, if you're Red Cross, they might be here for all the preparedness and run up to hurricane season but after that they pop smoke and they head north. I don't blame 'em, I would too.

Snowbirds, seasonal residents who generally spend the winter months in Florida and summer months in the northern states or Canada, have been found to play an important factor in the response to hurricanes (Montz and Tobin 2005). In one hand, fewer occupants in the summer months can ease recovery efforts, as there are fewer people to take care of. In addition, snowbirds can provide outside resources from the disaster site. Yet, lack of exposure to hazards events have the potential to make preparedness efforts more difficult, due to less volunteers and people with previous disaster preparedness and recovery experiences.

7.3.4 Orange

For this focus group, several themes arose from the analysis of the transcripts including: undocumented migrants, personal accountability, and mistrust of government. In discussing the challenges to resilience, one participant stated issues on the lack of trust affecting relations and recovery processes with undocumented migrants, particularly those working in agriculture:

> Well I work with a lot of, um, volunteer organizations and faith-based organizations and the challenges that we've had in the past, and we know that they'll be there for the future, are that there is segments of our community that really don't want to be recognized, and they're very apprehensive about government, and how government does work. So, we have some work to do, trying to get these people to realize that we're just there to help them, and to trust us. I'm talking specifically, about, sometimes; its illegal aliens, that are here, and they don't want to go to a shelter, because they think that somehow we're capturing their data. And uh, they won't come forward when they need assistance. We had a small tornado event a couple of years ago in the Apopka area, and the primary people up there effect were, um, local people that come and do harvesting at crop time. And a lot of them didn't even want to come forward and say that they needed help, that they had lost their belongings that they had lost their home site and everything else, and it was really hard. So we try to work through some of the faith-based organizations and things that are already in place to try to communicate through that organization and to the individuals that we're there just to help, and we're not there to, you know, round them up, or whatever the case may be. So that's a big barrier we face with trying to get help to the right place.

Indeed, several reports after Hurricane Katrina noted the misinformation and general mistrust of governmental officials as barriers to assist and give aids to immigrant communities (Blazer and Murphy 2008). The mistrust of government was not perceived as an issue just for undocumented migrants, but for the general public as well. As one respondent shared: "I think there's a general mistrust of government until something bad happens and then everybody turns to government at one time. And I think in the last few years it's gotten worse too. Because you know, it's like kick government time. And it has been for a while. You know, and it's been across the country, and it's until something happens and then they go 'please, help us!'" As with immigrant communities, distrust of government officials is a challenge as these agencies are responsible for emergency management. One example is seen the poor response by governmental officials after the aftermath of Hurricane Katrina (Nicholls and Picou 2013; Tierney 2012). In addition, the quality of the emergency management personnel and departments are contingent upon "the extent to which a local government has adopted policies to prepare for emergencies, mitigate their impacts, ensure an effective emergency response, and facilitate community recovery" (Henstra 2010, p. 238).

When asked about other issues, one responded perceived, in addition to the other issues discussed, there was a sense of personal accountability that has been lost and makes it difficult to build disaster resilience. In detail: "I think personal accountability, because, like they said, because, people, they think that the government should be taking care of them when something like this happens you know it's 'what do you mean you don't have water for us?' 'what do you mean I have to drive, you know, ten miles to get water?' and you know, you're like 'well why didn't you have water?'".

7.3.5 Osceola

In this county, the need for coordination between local government agencies was viewed as a challenge to the path to resilience. As one participant stated:

> In fact, our most recent event kind of highlighted that overall we don't have – there was no feel of overall coordination of what was going on between jurisdictions. As far as event he status up dates, I mean what's happening? Do you need any help? What's going on there, maybe we've got a crew in that area we can help you out with. Um so from that perspective there wasn't that. And it goes further than just local. Our, our water management system is not maintained locally. It's maintained our of West Palm Beach and there was no communication coming from WPB in the last event. And that is something that we are addressing now. It has been better in the past; it has been worse these past couple of years. Especially during the October event. I think there are a lot of things that we need to work out. And just the flow of updated information, you know, sometimes you can't mitigate for every circumstance, and you can't you know, um, prevent what's going to happen but you at least need to be aware what the possibilities are and what the worst case scenario is. And when you can't even get that, how do you plan to recover? Or how do you plan to be resilient? So that, that's one of the hard things from our perspective is you know, when you see a hurricane coming in from off shore, you know what worst case scenario could be. But when it picks up and rains for 10 days in a row, or rains for 2 full days an you can't get information of you know what's another inch of rain going to do. How, how far are we going to go under. What do we need to do? Cause you know the generally population out there, it hasn't rained in 3 days, and they think its fine. What they don't realize is that the water levels are so far up there that with a half inch of rain, we'll all be under water. And so that's the perception of the public, that there's not a problems. But there is a reality that there really could be. So its getting the best information available so we don't panic the public, but we're ready if we need to be ready. And so recovering from that and uh, in making sure that it's mitigated for is important but we need the information. The information is key.

Another participant added:

> Communication amongst the agencies and all, I think there has been a lot of effort to improve that. Um, I don't really know how to gage it, other than I know that I have attended more meetings regarding that in the past couple years than I ever remember. So I know that the effort is being made. Uh, I don't know how successful it is. I don't know how to gage that. As far as challenges for our organization, we continue to have the challenge that our service area is the entire county and regions of Polk. And, we're dependent on power from a number of different power companies, as there are a number of power companies that serve those area and as a utility with over 300 lift stations and over 20 facilities scattered about both Osceola and Polk county uh, maintaining power during disaster situations or, um, we're entirely dependent on that except for areas where we have emergency generators. And you can't cover the entire area with emergency generators. And even if you could you couldn't service them in the even of a disaster. At least not for very long. So we're greatly dependent in disaster situations on power and our coordination with the power utilities to maintain that and to get that back in a prioritized manner. So that is a challenge and I know that in some events that we're had in the past that it was difficult to even get representatives from the power company in the EOC during certain events. And that I think still continues to be a problem. Everybody seems to be doing their own thing. [Pauses] That is one of the biggest challenges that we have.

Communication and coordination among agencies are critical aspects to disaster resilience. Coordination consists of synchronizing a community's resources in terms of finances, facilities and volunteers. A more detailed definition consists of coordination being the "degree to which there are adequate networks among the organizational parts for intraorganizational communication" (Kapucu 2006, p. 209). There is a goal of sharing information and being able to, more effectively, mitigate disaster response and recovery (Comfort 2007; Kapucu 2006). Moreover, coordinated efforts allow for a common operating picture, which ties into FEMA's desire for a whole community approach. With each disaster event causing the need for complex partnerships, as they do not respect borders, coordination becomes essential to quicken the decision making process (Tierney 2012).

Some researchers deem coordination one of the most important disaster response and recovery aspects as the extent of impact from a disaster or hazard is unpredictable. For many situations, there is a need for officials from federal, state and local governments to come together. Furthermore, there can be confusion as to the roles and responsibilities of each sector. "Community response to a disaster or emergency can be successful if capacity building has taken place and the collaborating agencies have developed a shared vision [and] a common understanding of the problem" (Kapucu et al. 2013, p. 3). Citizens rarely have specific requests from emergency managers so it is vital for recovery and response efforts to be as comprehensive as possible (Henstra 2010).

7.3.6 Seminole

In this county, participant's expressed disaster resilience issues with the training of volunteers and other emergency management personnel. One of the focus groups participants stated:

> I think there is a lack of training in a lot of these departments, that they don't do until zero-hour and then they want to know "what are we supposed to do, and how are we supposed to do it and I think that goes with our volunteers too. Volunteers are wonderful; we have about four volunteers that we know we can depend on to come in here. And they want to help, however they are very, um, elderly, so you know they're limited. You're limited in the things you can give volunteers, uh, you've got to look at everything whenever you're dishing things out. But I think training is a big issue. And people don't come to training. We offer training constantly and people don't come until there's either an emergency or they're forced to come.

The inclusion and training of volunteers is viewed as a necessary tool for disaster resilient communities (Henstra 2010). It is not uncommon to have individuals converge at a disaster site to assist in the response and recovery efforts. The volunteers become an integral part of the response efforts as they assist in distributing supplies, reporting damages, clearing debris, and conveying information to other community members (Henstra 2010). Indeed, federal frameworks, such as the "whole community" where all sectors of the community (business, non-profit groups, citizen groups, and governmental officials), are critical pieces to disaster resilience (Edwards 2013) and incorporate how organizations and individual members, such as community emergency response teams, are able to be mobilized (Henstra 2010). Thus, training becomes essential as volunteers, although viewed as helpful in disaster situations, sometimes can become a liability or a hindrance if not properly trained (Barsky et al. 2007).

7.3.7 Sumter

Finally, in this rural county space between people was viewed as a particular challenge to disaster resilience. One participant stated:

> In a rural community like ours, you've got space between people. The community is fragmented. When you get outside of the Villages, which is more of a metropolitan or suburban type of environment, when you get further out past say Wildwood, then you're talking about people living maybe on two or three acres of land or maybe their neighbor, the next house, is perhaps a half a mile away. Uh, I think it's hard to get them all focused on the same thing, to communicate and idea perhaps, and to get them, to uh, come together, as you know. As No. 6 was saying, in the city, when something happens in one part of the town, everyone knows it. They may not feel they're affected by it, but they're aware of what's just happened. And, uh, here, something could happen down in the Croom (reference to local area) and people up north, unless they're turning on the TV station that's going to tell 'em about it, they don't even know what's going on down there.

Another added:

> Again, the space between people and the ability to notify other people that there is something wrong becomes a problem in a rural area where, you know, in a very close neighborhood in a large city, you just throw the window open and say 'Charlie! I need your help!' but, you know if the telephones are out and the power grid is down and everything, people can't get on the phone. They can't call people and say, you know, the house just got blown over, or something like that, because there are no services. So, it becomes a problem of communications and being able to get the regular communications back up and running, but also until that occurs, it depends on being able to communicate by remote means, like radio and things like that.

As a rural county, the issues of space are perceived as a challenge to disaster resilience. Indeed, as discussed in Chap. 5, there appears to be a perception of rural areas being more susceptible to disasters and hazards with an inability to recover as quickly as more urban areas (Durant 2011). Moreover, there are considerable geographical issues, including the distance between rural and urban communities (Kapucu et al. 2013).

7.4 Conclusion

Some of the ways to increase (and sometimes create) a community's resilience capability consists of understanding the interconnection of a community's assets and investments in future projection. In also involves knowledge of the vulnerabilities of a community and their disaster hazard risks. In defining resilience, we found barriers to increasing a community's capacity linked to aspects such as socioeconomic, complacency, shifting demographics and resources in terms of shelters and facilities. These findings have been supported through personal encounters by focus group participants in Brevard, Lake, Orange, Osceola, Sumter, and Seminole Counties. As we move from over-encompassing elements of disaster resilience it becomes increasingly importance to take into account the perceptions of disaster management personnel who are at the forefront and have an intimate knowledge of the obstacles and challenges to disaster resilience. Furthermore, models like the Adaptive Resilience & Community Capital (Fig. 6.1) provide a framework to better understand the process by which communities are able to enhance their disaster resilience.

References

Barsky, L. E., Trainor, J. E., Torres, M. R., & Aguirre, B. (2007). Managing volunteers: FEMA's urban search and rescue program and interactions with unaffiliated responders in disaster response. *Disasters, 31*(4), 495–507.

Blazer, J. B., & Murphy, B. (2008). Addressing the needs of immigrants and limited English communities in disaster planning and relief. *Immigrants' Rights Update, 22*(8), 1–12.

Combaz, E. (2014). *Disaster resilience: Topic guide*. Birmingham: GSDRC, University of Birmingham.

Comfort, L. K. (2007). Crisis management in hindsight: Cognition, communication, coordination, and control. *Public Administration Review, 67*(S1), 189–197.

Deyle, R. E., & Smith, R. A. (1998). Local government compliance with state planning mandates: The effects of state implementation in Florida. *Journal of the American Planning Association, 64*(4), 457–469.

Durant, T. J. (2011). The utility of vulnerability and social capital theories in studying the impact of Hurricane Katrina on the elderly. *Journal of Family Issues, 32*(10), 1285–1302.

Edwards, F. (2013). All hazards, whole community: Creating resiliency. In N. Kapucu, C. Hawkins, & F. Rivera (Eds.), *Disaster resiliency: Interdisciplinary perspectives* (pp. 21–48). New York: Taylor & Francis/Routledge.

Henstra, D. (2010). Evaluating local government emergency management programs: What framework should public managers adopt? *Public Administration Review, 70*(2), 236–246.

Kapucu, N. (2006). Interagency communication networks during emergencies: Boundary spanners in multi-agency coordination. *The American Review of Public Administration, 36*(2), 207–225.

Kapucu, N., Hawkins, C., & Rivera, F. (2013). Emerging research in disaster resiliency and sustainability: Implications for policy and practice. In N. Kapucu, C. Hawkins, & F. Rivera (Eds.), *Disaster resiliency: Interdisciplinary perspectives* (pp. 355–358). New York: Routledge.

Montz, B. E., & Tobin, G. A. (2005). *Snowbirds and senior living developments: An analysis of vulnerability associated with Hurricane Charley*. Boulder: Natural Hazards Research and Applications Information Center.

National Academy of Sciences (NAS). (2012). *Disaster resilience: A national imperative*. Washington, DC: The National Academies Press.

Nicholls, K., & Picou, J. S. (2013). The impact of Hurricane Katrina on trust in government. *Social Science Quarterly, 94*(2), 344–361.

O'Sullivan, M. (2003). The fundamental attribution error in detecting deception: The boy-who-cried-wolf effect. *Personality and Social Psychology Bulletin, 29*(10), 1316–1327.

Rivera, F. I., & Settembrino, M. R. (2013). Chapter 3: Sociological insights on the role of social capital in disaster resilience. In N. Kapucu, C. Hawkins, & F. Rivera (Eds.), *Disaster resiliency: Interdisciplinary perspectives* (pp. 48–60). New York: Routledge.

Tierney, K. (2012). Disaster governance: Social, political, and economic dimensions. *Annual Review of Environment and Resources, 37*, 341–363.

Tobin, G. A. (1999). Sustainability and community resilience: The holy grail of hazards planning? *Global Environmental Change Part B: Environmental Hazards, 1*(1), 13–25.

Wang, X., & Kapucu, N. (2008). Public complacency under repeated emergency threats: Some empirical evidence. *Journal of Public Administration Research and Theory, 18*(1), 57–78.

Chapter 8
Communicating Resilience

Abstract This chapter explores the importance of understanding how social aspects like communication and social capital impact community resilience and influence the various public, private, and non-profit agencies that respond to crises. It also explores the perceptions of emergency management personnel in communicating with the public during a disaster situation.

Keywords Communication • Resilience • Social capital • Communication tools • Technology • Social media • Florida

8.1 Introduction

Effective communication is important during each phase of emergency management. In the mitigation and preparedness phases, communication helps to exchange and confirm information and efforts to be prepared by appropriate stakeholders and organizations prior to an event. During the response and recovery phases, effective communication helps people do the right thing during and after a disaster so they have the best bet of surviving with as little loss as possible. However, communication is not effective in these phases unless it is a two-way process to have input and participation from both the sender and receiver of information. Effective communication has a multitude of components to work in all four phases of emergency management. First and foremost, effective communication requires active listening to ensure understanding of the message and to respond appropriately to the information which could save lives. Of course, to ensure your audience is effectively listening, the speaker must recognize the audience and tailor the message for understanding. Knowing who the audience is in a situation can not only enable the speaker to adjust the message, but also his/her posture, demeanor, and body language for even better message sending. Unfortunately, the decision makers and media did not fully grasp these communication techniques during the 2004 hurricane season in Florida which caused an epidemic of complacency.

© Springer International Publishing Switzerland 2015   97
F.I. Rivera, N. Kapucu, *Disaster Vulnerability, Hazards and Resilience*,
Environmental Hazards, DOI 10.1007/978-3-319-16453-3_8

8.2   Complacency: First Hand Experience Florida Hurricane Season 2004

During the 2004 Hurricane Season, four hurricanes appeared within about a month timeframe from each other: Charley, Frances, Ivan, and Jeanne. Charley directly affect affected the Western coasts of Florida where it caused power outages, destructions of buildings, and about two dozen deaths. Frances was a Category 4 and caused severe damage to eastern Florida counties leaving millions without electricity. Ivan mostly hit the panhandle while Jeanne weaken over Florida but still caused several million in damage and left millions without electricity. These same four hurricanes, in addition to a multitude of severe tropical storms of that season, devastated the Bahamas, Jamaica, and Haiti. These islands were flooded, had major infrastructure destroyed, and no power for exceptionally long periods of time. In Florida, infrastructure is built with better materials to endure more havoc and emergency response and management received far more funding then the island neighbors to the south of Florida.

Florida is accustomed to hurricanes. In fact, when there were hurricanes, several families would invite families or neighbors over for hurricane parties. People stocked up food, beverages, candles, flashlights, and batteries and would just enjoy the time together without work or school. Even when power is lost, some see it lovely to have the house filled with candles. After the hurricanes passed, people would go out with the neighbors to assess the damage together.

The contradiction of results between the islands and nations of the south to Florida helps understand how complacency occurs before, during, and after a disaster occurs. Complacency is a feeling of security when unaware of potential danger. Kapucu and Özerdem (2013) discuss the Florida population's complacency during the 2004 hurricane season and how the preparations taken after the first warning gave the public confidence not to prepare further for subsequent disasters. That season had repeated hurricanes and tropical storms within a short timeframe and the warnings felt giant compared to the less than devastating results of the disasters. Wang and Kapucu (2008) recommend decision makers deliver detailed information about how to prepare for the specific disaster and do so in a timely manner to prevent complacency and allow the public to effectively protect themselves in the emergency management process (For more on complacency see Chap. 6).

Risk communication is another important factor in enhancing community resilience and should involve the development of relaying specific messages for both officials and the public at large, which are at risk from impact from disasters. Risk communication not only deals with immediate problems and crisis situations, but also seeking ways to adapt, and dedicate energy toward new ways to disseminate information to all shareholders involved. One should focus on providing renewal in reforming failed strategies associated with risk management. This renewal also is focused on increasing significant resources to help transition into working toward achieving new and important objectives that is geared towards empowering all shareholders in rescue and recovery efforts.

All of this is important to consider when dealing with emergency management and disaster response shareholders and the public. The concept of shareholder cooperation is also vital in risk communication. They, being the public can be valuable partners in risk response. However, any emergency manager must make sure the public has prompt access to information concerning the disaster and ways to get through it without injury or harm, or even to potentially save lives. All of this can be utilized through a crisis management plan, which guides all shareholders through preventing a risk from escalating or to mitigate and properly respond to the effects of a disaster when it emerges. This recent outbreak of Ebola in Texas and in Africa could very well provide more examples of how to properly communicate through emergency channels and its stakeholders. The situation seems to be direr as time passes, and indeed we will see how these principles discussed here are executed.

Garnett and Kouzmin (2007) highlight Hurricane Katrina as one of the largest communication disaster; with communication gaps, missed signals, information technology failures, administrative buffering, and misinterpretations that delayed the crisis recognition and the response to the devastation. They highlighted four aspects of crisis communication; interpersonal influence, media relations, technology showcase, and crisis communication as interorganizational networking. Hurricane Katrina demonstrated failures with operability and interoperability of communications technology, which have then prevented or delayed effective preparedness, response, and recovery.

Wang and Kapucu (2008) provide several effective communication strategies based off their research these include; monitoring public complacency in pre-disaster preparation (important for effective responses), developing effective communication strategies to deliver the most accurate information in a timely manner to the public during emergencies, the need to include several styles of information delivery including visual images and information tailored to specific residential groups, and realizing that the public response to warning is not a simple stimulus-response reaction. In addition, they emphasize on the importance of partnering with the local and national media to provide accurate and timely information to the public. Effective partnerships with media help emergency management professions research the general public easily to help them fight complacency and provide them with critical information.

8.3   Relations with News Media

As previously stated, effective partnerships with the news media have the potential to help emergency management personnel reach the general public easily and provide them with critical information (Wang and Kapucu 2008). Indeed, a participant from the Orange County focus group stated:

> We have meetings with the news media; our communications manager meets with them on a regular basis. When we were doing the Emergency Alert System (EAS) test recently, a nationwide test, we had several face-to-face meetings, we provided lots of information, had

lots of dialogue. Of course the EAS test didn't go as well as they had hoped, but that was the whole point of putting it through a test. You know, it wasn't a case of figuring out if it's a success or not, no it was a success because it did what it was supposed to do. It identified the issues that need to be worked on. So I think, I think that in general, we've gotten better as far as having that relationship with the news media and the information that's going out in most cases is becoming more reliable.

There was a recognition that working with the news media is an important source for communicating and providing accurate information to the public. Nonetheless, there were various concerns about some aspects of the news media cycle, particularly the need for ratings and sensationalism.

As one participant from Lake County shared: "I think one of the challenges (in dealing with the news media) and they've kind of touched on that also, is in the world of 24 hour news, it's less about factual reporting and providing information and more about tracking viewership and hype. So the hype side of things, you can get people involved or over involved, which is sometimes good – but the hype side also is, sort of like the example that was just given about instead of just reporting on the factual side of things, that hype of "you guys didn't bother responding." So that's a big challenge in that eating up a lot of time just answering all that because they're trying to drum up some kind of story to grab viewers as opposed to just providing information. You might want to get information put out that, so you provide that to them, but it isn't cool or sexy enough to put it on television so that's one of the challenges…" A similar sentiment was shared by one Orange County focus group participant sharing an experience from the 2004 hurricane season. In detail,

"In 2004 they put a list in the newspaper, a whole page listing that these are all of the shelters. What they did is they got a copy of all the schools from the school board and then threw them on the newspaper which doesn't necessarily mean that all of those schools are open as shelters, and it gives a misperception as to what is really available to the public because all of that information was wrong. I was sitting next to Emergency Support Function 14 (ESF 14-Long Term Community Recovery) at the time, and there were plenty of times, ESF 14 is a PIO, public information officer, and they're the ones that are supposed to release the information. We'd be sitting there watching TV and we'd see something that was like totally, obviously wrong on TV. And it's like "where did they get that from?" And it was like they (news media) just picked something up out of the air and if they can't get something from us they make it up. And they just generate whatever information they want out there, you know, because they all want to be number 1. (Group members affirm) You know they all want to be number one for ratings and so they're going to sensationalize everything when it might not even be the truth…".

Regarding the need for ratings and viewership a Seminole County focus group participant stated: "The media its self has a problem, and the problem is they want numbers. So they think a disaster is lots of coverage. If they just come and report it, and put the information out about where there is transportation difficulties or something like that, and then leave it alone. Everybody would be so much happier. Because they really feed the anger or those people that are sitting there, watching television on a battery television set, about how great the power outage is, and how slow the power company is to respond. Which isn't a true fact. I think that they, the over dramatization that the news coverage, that the news media tends to give, just to build their own numbers really needs to change".

8.3.1 Worked Against Us

Sometimes miscommunication with the news media can have serious unintended consequences, as shared by one Lake County focus group participant:

I'll give you one example of how media worked against us in an emergency management role was that all the reporters that we had in the field were just like sponges trying to get any story they could. The very next day after the impact of Charley the news media reported on those that were getting electricity by generator and things like that, was that you better keep your generators in a secure location because we're now getting reports from the sheriff's department that people are stealing generators. Well, you know, generators burn fossil fuels with is CO_2 (Carbon Dioxide), so if you hear that, you know "put your generator in a secure place" they were bringing them into their garage and closing the door. And that first day we had 32 critical CO_2 poisonings. And what we had to do was switch over and drive the streets at night, you know, listening for the hum of the generator or lights. And we had an incident early in the school year where two grand parents had passed away from accidentally leaving their Cadillac in their garage and our crews didn't pick up on it, and we got them out of there. But what we ended up doing was buying CO_2 detectors for each one of our first out units which was invaluable when Charley came in August because we were able to find a family of 5, and the father was the one, the children were unconscious, the mother was unconscious and the father was groggy and before they even got into the house the CO_2 detectors were going off. All lived, one of the children did have some deficits from it but that was how the media really worked against us. They reported without thinking about the CO_2 danger. So, it can work against you as well. Then it was like crisis mode. Trying to get the word out not to put them inside. Chain them to the tree if you have to.

Aware of the potential for miscommunication a focus group participant from Seminole County shared the following advice when dealing with the news media: "As far as dealing with the media, you need to control the media. You need to be able to get your message out and use them to get the message out to the populous and the public at large. So you need to, I don't want to say manipulate them, but give them, feed them the message that you want to get out. Not their message about sensationalism, as mentioned before. You need to, uh, funnel them, with your messaging track and your message to the public of what they need to do or not need to do, or where to go or where not to go and so on and so worth. So you need to use them as a tool, or a sub-agent, as a sub-contractor of you, to get that message out. So you can't let them manipulate you, you have to funnel and focus them.

8.4 Communication and Sensemaking

A consistent attribute of social response to crises is communication. This cyclical process of intake and output not only impacts decision-making process, but it affects the ability for a responder to actually comprehend the situation. Sensemaking occurs when individuals rationalize information and apply meaning to it (Weick 1993). When dealing with a crisis, the incoming information can be too much to process resulting in lapse judgment and various consequences. To circumvent these issues, creating a common operating picture can help streamline information and positively improve communication among the diverse and numerous aid organizations and agencies (Wolbers and Boersma 2013).

The common operating picture (COP) is considered one of the most significant solutions in emergency management (Wolbers and Boersma 2013). The COP is considered a tool to improve the quality of information as well as support situational awareness development. Using geographic visuals and checklists the COP can be articulated to invested stakeholders. Although the COP is useful, it is important for the stakeholders to understand whether the document is informational or is a plan of action. Once the COP is articulated to the various stakeholders, issues in sensemaking can be reduced.

In 1993, Weick published an analysis of the Mann Gulch Disaster and responders' failure to recognize the reality of the situation they were in. He points the loss of lives was an unfortunate consequence of sensemaking collapse and disintegration of the organization's structure. For example, the responding crew was expecting a 10:00 fire, a category depicting response time, yet they encountered a more serious situation and were unable to evacuate before it was too late.

Similar to the Mann Gulch disaster, management failure occurred in response to Hurricane Katrina due to lack of sensemaking and cognition of administrators (Comfort 2007). The scope of the hurricane exceeded the capacity of public organizations, at every level, and incited input from private and nonprofit organizations. This situation led to an increase in number and diversity of response organizations, which also caused dissolution of the common operating picture. This resulted in hierarchy becoming the dominant management perspective and variations of information transferred among organizations (Comfort 2007). The disjointed responses undermined the healthy communication, coordination, and control of responders, since they were unable to access accurate information in a timely manner.

Human cognition, or comprehension, can be discussed on two levels: individual and social. Individual level cognition refers to a person's capacity. Social, or distributed, cognition not only incorporates individuals, but also their social environment and related artifacts. In disaster management settings, two components of distributed cognition are fulfilled by stakeholders, or social groups, and information and communication technology (ICT), or artifacts (Celik and Corbacioglu 2010). First, organizations use decision support systems to make accurate and reliable decisions before, during, and after an event. These systems include inputs from other organizations, technological capacities for information processing, and sharing this information for appropriate users. Second, ICT provides artifacts for emergency responders to develop a cognitive process to monitor threats or change conditions in their communities. The timely and congruent communication is vital for successful coordination and control of emergency management efforts. Comfort (2008) gives an example of distributed cognition in the evacuations for Hurricane Katrina:

[Evacuations] would have meant alerting the school bus drivers on Friday to the potential need for their services on the weekend, given the high probability of activating the evacuation plan for the City of New Orleans as protection from the impending storm. Decisions occur in social settings, and no single individual is able to conduct a successful evacuation of a city of 450,000 without the coordinated actions of the mayor, the school bus drivers, the garage mechanics, the City Police, the State Police, and the evacuees themselves (p. 5).

For a management system to be successful and increase the resiliency of a community, information processing needs to include four decision-making subsets. The first is detecting the risk. Second, the system must interpret the risk within its immediate context. Third, the risk needs to be communicated to various stakeholders within a wide region. Last, the responders need to mobilize and organize themselves for collective action to reduce risk (Comfort et al. 2010). These subsets build upon each other and attribute to the development of resiliency.

For the State of Florida, the Division of Emergency Management operates an 24 h crisis line at the State Emergency Operation Center to maintain constant information flow regardless of if the state is experiencing a disaster or hazard (Division of Emergency Management 2014a). The technology involved includes telephone systems, satellites and radios, and weather predictors. Moreover, these information systems are connected to various federal and local emergency management systems to maintain the information network. Moreover, the communications unit for the Division of Emergency Management (2014b) is responsible for maintenance of the network and developing the capability for communication to mitigate any potential issues.

An example of information and communication technology development is through the state's Emergency Communications Number E911 State Plan Act. This act mandates the Department of Management Services to develop and update their communications plans for quality service provision. The mandates, more specifically, include:

- The roles, responsibilities, and requirements of the public agency emergency communications system for each entity of local government in Florida.
- A system, designed to meet specific local government requirements for public emergency communications agencies, which shall include law enforcement, firefighting, and emergency medical services and may include other emergency services such as poison control, suicide prevention, and emergency management services.
- Identification of interagency coordination and mutual aid agreements necessary to develop an effective E911 system.
- A funding provision that identifies the costs necessary to implement the E911 system (Division of Telecommunications 2010).

Another useful guide to improve communication is through the National Emergency Communications Plan. This document was established to help the strategic planning for local, state, and tribal emergency management personnel. The overall vision incorporates the following seven objectives:

1. Formal decision-making structures and clearly defined leadership roles coordinate emergency communications capabilities.
2. Federal emergency communications programs and initiatives are collaborative across agencies and aligned to achieve national goals.
3. Emergency responders employ common planning and operational protocols to effectively use their resources and personnel.
4. Emerging technologies are integrated with current emergency communications capabilities through standards implementation, research and development, and testing and evaluation.
5. Emergency responders have shared approaches to training and exercises, improved technical expertise, and enhanced response capabilities.
6. All levels of government drive long-term advancements in emergency communications through integrated strategic planning procedures, appropriate resource allocations, and public-private partnerships.
7. The Nation has integrated preparedness, mitigation, response, and recovery capabilities to communicate during significant events (Department of Homeland Security 2008, p. 2).

Alongside the national document, the Federal Emergency Management Agency (2014) incorporates an independent study course focused on communication. Participants within this course will not only develop basic communication skills, but will help individuals learn: (a) how to communicate in an emergency, (b) how to identify community-specific communication issues, (c) use technology as a communication tool, (d) develop effective oral communication, and (e) how to prepare for an oral presentation (FEMA 2014).

8.5 Communication Strategies

Effective communication can affect people's perception of risk. It is imperative to have sound public relations with the community one is serving in a natural disaster or emergency situation. With the correct use of Public Information Officers (PIOs), the public can be inspired, have trust in government services and stimulate dialogue with elected officials for the proper emergency procedures before, during, and after a disaster (Kapucu and Özerdem 2013). If the public and a community feel that the risk is too much or too low (risk misconception), then the emergency management systems in that particular locality, or aspect, will suffer negative consequences. Another aspect of risk perception is risk assessment. After a natural disaster or an emergency situation, emergency management organizations with stakeholders must engage in thorough evaluation of the emergency situation and what were the actions taken. Correct information dissemination is key in risk assessment and it deals with the proper information given at the proper time to the proper recipients. Information dissemination is a scientific art form that takes many specialists to perform, but is a necessity in emergency management (Kapucu and Özerdem 2013).

Risk communication during an emergency or disaster requires a thorough developing strategy before a disaster strikes. Effective communication development at this stage is key. In emergency management and planning, a strategy for communication can be divided into two categories:

1. Internal communications, or simply put – communications of commands, requests, directions and responses by and between emergency management and response teams and other personnel, associated with disaster response.
2. External communications, or communicating notifications, warnings, and general information to the public through news outlets and social media.

It seems to me that through these categories, local emergency service organizations focuses on lifesaving strategies that works toward reestablishing control in the disaster area. Any county or municipality emergency management agency will become the central point of coordination and control. Any proper available form of notification to provide information in a timely manner is to be used during an emergency situation. This flow of information may run through these two channels of internal and external communications.

There are other factors that all emergency managers should also consider when allowing these two information channels to flow properly. The reports that first become available from the epic center of the damaged area may be fragmented and provide an incomplete picture of damage and loss of life. Weather and other environmental factors and even terrorist related threats (such as 9/11) could also restrict communications such as cellphones and other transportable means of communication in a hit area. Emergency communications systems in the damaged area may become overwhelmed or even inoperable during a disaster or in the aftermath of a disaster.

There are a number of ways to assess these situations. First, any good emergency manager should make sure they ideally develop and maintain a Continuity of Operations Plan (COOP) to ensure uninterrupted operations during disasters and this should include communications. In this COOP, emergency managers are to develop and maintain a training plan for personnel with the latest in communications technologies. In this process, they should identify critical infrastructure necessary to maintain proper communications internally. Any emergency manager should be able to gather information from an impacted area and determine which communications systems works, including cellular networks, land-lines, and even HAM radios. One should assess the communications requirements for damage assessment on an impacted area while coordinating the gathering and distribution of the communication equipment. Afterwards, an emergency response team (and manager included) should also mobilize resources and communications support with other authorities such as the police, fire department, other government agencies, and volunteer agencies as requested. This entire process is more of a basic sketch on how communication systems in emergency management may occur.

The internal processes of communication in emergency management requires, first, to collect information, such as identifying those info sources and their availability. This first step also involves an "assessment of the integrity or truthfulness of information sources to filter out possible rumors or misinformation" (Kapucu et al. 2008, p. 172). The next step is to process information strategies to identify information useful for "decision making" (p. 173). The last step is listed as "information dissemination and exchange that emphasizes the modes of information delivery…" (Kapucu et al. 2008, p. 173).

Having an effective and efficient communication strategy during an emergency situation requires developing these strategies in advance, practicing this strategy and constantly developing and building upon it. Communication strategies provide for a structure to increase coordination and knowledge across agencies involved and also to better inform and educate the general public. One place to start when looking to develop a communication strategy is the Federal Emergency Management Agency (FEMA) National Emergency Communication Plan. This plan identifies the capabilities needed by emergency responders to ensure the availability and interoperability of communication during emergencies, recommend both short and long term solutions for ensuring interoperability, set goals and time frames for deployment of interoperable emergency communication systems and to set dates for which Federal agencies, and State, and local governments expect to achieve baseline level of national interoperable communications. The plans vision is to "ensure interoperability and continuity of communications to allow emergency responders to communicate as needed, on demand and as authorize at all levels of government and across all disciplines" (DHS 2008). DHS's National Emergency Communications Plan is a good place to start when developing a communication strategy other things to consider are how to communicate effectively with the general public. When considering communications externally with the public it is important as the text states to have good partnerships with the local and national media.

New communication technologies can assist in emergency information collection and dissemination. Latonero and Shklovski (2011) point out that social media platforms such as Twitter "gave individuals the unprecedented ability to rapidly broadcast and exchange small amounts of information with large audiences regard less of distance" (p. 2). They, however, point out that currently very few official government organizations use Twitter and other social media tools. They also claim that "social media provides the potential for interactive, participatory, synchronic, two-way communication" (p. 6). The external communication can be enhanced by the usage of social media. Emergency managers should be cognizant of both streams of communication since people's lives and livelihood are at stake when disasters impact them directly. Kapucu and Özerdem (2013) stressed that modern "technologies at the disposal of emergency management professionals are much more sophisticated than the tools available to practitioners in earlier areas" (p. 179). Indeed, utilizing all tools available to disseminate information through proper channels is vital to the emergency management professional.

8.6   Communication Technology

Within the process of creating a management system is information communication technology. With the complexity of crises, it is critical for administrators to have access and have the ability to efficiently disperse information. Through the knowledge, responders are able to generate a common operating picture. Therefore, information communication technology (ICT) is critical to support effective decision-making under complex and uncertain disaster conditions (Celik and Corbacioglu 2010). ICT also enhances cognitive capacity of emergency managers and enables them in processing large volumes of information in short time periods (Comfort 2007).

In regards to information processing, Comfort and her colleagues propose a bowtie model within a disaster management system (Comfort 2007; Comfort et al. 2010). The model consists of three integrated parts: (1) data collection, (2) data analysis, and (3) organizational action (see Fig. 8.1). Data collection involves various contributors of the decision-making system, such as weather services who predict storms or the American Red Cross who provides shelters. Data analysis requires

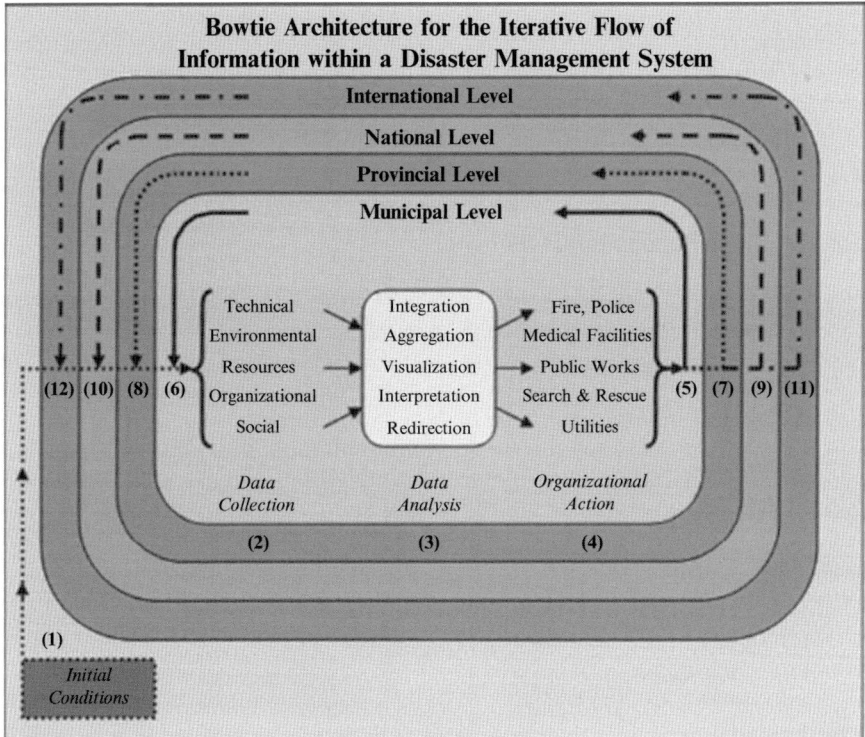

Fig. 8.1 Bowtie model for information system (Comfort 2007)

various expertise that may not necessarily be prevalent in one organization. Organizational action is critical for risk communication, as data partners are informed about the conditions and the risks of the situation. The information results in organizations structuring their actions accordingly and communicating potential risks to the wider public (O'Brien and O'Keefe 2010; Paton 2007). This, in turn, increases the community's resiliency.

Although the creation of models is important for increasing the flow of information and improving communication systems, the practical implementation is imperative as it allows for potential issues to surface. For example, it is imperative to maintain a connection without incurring any failures (Kapucu 2006). During September 11, 2001, the technological infrastructure of New York City was seriously damaged. These damages consisted of:

> Complete or partial loss of five smaller buildings in the immediate area and heavy damage to other buildings in the area. In addition, the electrical power generation and distribution system for lower Manhattan was destroyed; the water distribution system, dependent on electricity for pumping water, was disabled; gas pipelines were heavily damaged; and the telephone and telecommunications services were seriously disrupted (Kapucu 2006, p. 208).

The damage to the infrastructure resulted in emergency management personnel scrambling not only to initiate response activities related to the needs of affected individuals, but to also create avenues for communication to create a common operating picture. Kapucu (2006) analyzed the response to September 11th and generated a model of interorganizational communication and coordination to implement in crisis situations (see Fig. 8.2). The model provides a visual of how negative impacts of an event can be mitigated through sharing information, placing importance on information technology, to increase communication and provide better services to the public.

Another practical exemplar application related to the emergency management phases of response and recovery is Google's Person Finder. This application was created in response to the Haiti Earthquake to help individuals locate other people or provide information about a missing person. After the Haiti Earthquake, Google's

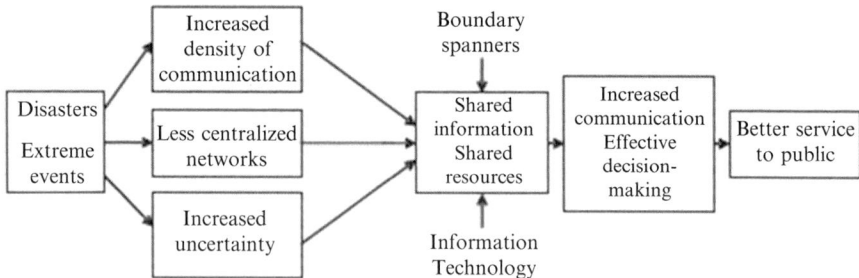

Fig. 8.2 Communication in emergencies (Kapucu 2006)

Person Finder was utilized after disaster events in Chile and Japan. Through the creation of an information network, individuals were able to obtain some control over the effects of the disaster.

Regarding the State of Florida, in 2008, the Army National Guard implemented a Regional Emergency Response Network with quality, state-of-the-art technology for disaster response (Boland 2008). The network system included radios with higher capabilities due to issues during the 2004 and 2005 hurricane seasons. "The network expands the Guard's presence in austere operating environments and provides reachback capabilities through satellite technology. It also offers radio suite capability that is tied into Florida's law enforcement radio system. And unlike other communications solutions, the new network is highly mobile" (Boland 2008, para. 5). These improvements not only developed the response technology for the guard, but also increased the social capital with other responding organizations and agencies.

8.7 Technology and Communication

Previously we discussed the different information systems currently available to disperse information such as the weather service, Google's Person Finder and other emergency response networks in addition to the traditional venues of television and the radio. Focus group participants share the views on the use of technology, particularly the internet, which we discussed below.

8.7.1 Internet

Most participants share the sentiment that the internet is a useful tool for sharing information with the public. Indeed, all of the Counties studied have a web-page portal that the public can access to find disaster related information. Indeed, Orange County focus group participants shared with enthusiasm the unveiling of a cellular phone app, where subscribers can access for weather related updates and information, in detail:

> We've got, we've now got an app, an iPhone app, to give information. (Other member: and android, android is live now) and so we've taking advantage of those sorts of things. We have been working with the National Weather Service and have gotten the authority to put information out there, as well go out on the EAS system, emergency alert system, that we can generate internally that will hit all of the media across the community. And, it'll go out on the weather radios. So we've got, now (we) have that capability, so we're trying to take advantage of all of the technology and I think we're doing' a pretty good job of at least staying atop of the technology. We're not on the leading edge, but we're at least keeping up with it.

8.8   Communication and Social Capital/Trust

Once risks have been perceived, interpreted, and communicated, it is time for the multitude of public, private, and nonprofit organizations and agencies to assist communities in responding and recovering. "To build collective resilience, communities must reduce risk and resource inequities, engage local people in mitigation, create organizational linkages, boost and protect social supports, and plan for not having a plan, which requires flexibility, decision-making skills, and trusted sources of information that function in the face of unknowns" (Norris et al. 2008). These supportive networks can be connected to a community's social capital. Social capital, although it varies in definition, there is a consistent incorporation of the cumulative actual or potential resources that mutually benefit social networks and are influenced by norms and trust (Aldrich and Meyer 2014).

By strengthening social capital, communities are able to increase their resilience (Kapucu 2006). Aldrich and Meyer (2014) propose building social capital, to increase resilience, through social events, incentives for volunteerism, and strategic planning for layout and architectural structures of every community. It also needs to be noted that building social capital is important, but also sustaining them (Kapucu and Garayev 2011). One way is through information and communication technology. If a communication network is found to positively impact social capital, it then, in turn, adds to the community's resiliency.

With social capital being an indicator of a community's resilience, collaborations become crucial (Johnson et al. 2013). Whether formal or informal, stakeholders must work towards strengthening their connections for the times when crises arise. If the leadership within organizations and agencies does not acknowledge the importance, then networks can be negatively impacted, which then negatively affects resilience. "Network-based social capital may be seen as integral to the resilience capacity of a community to respond quickly and flexibly during emergencies" (Johnson et al. 2013, p. 5). Therefore, leaders must develop their communication skills along with their professionalism, volunteer interactions, and managerial skills.

Skills focusing on communication are especially useful in response activities when dealing with volunteers. Volunteer management is a component heavily influenced by communication in regards to the numbers of individuals who come to help with relief efforts along with the number of organizations who offer their services as well. Within the State of Florida, Volunteer Florida is the lead agency for the volunteer emergency support function and helps the Florida Division of Emergency Management operations by coordinating volunteers while also assisting with donations (Volunteer Florida 2014).

Effective communication is a core factor that often determines the success of emergency management systems in all four phases; mitigation, *preparedness, response and recovery* (Garnett and Kouzmin 2007; Kapucu and Özerdem 2013; FEMA). Since most disasters are managed at the local level, one of the best management practices for local government is the development of a risk communication plan. A successful risk communication plan requires a commitment from the governing body to provide resources such as money, personnel, and training to develop and maintain the plan (Ng and Hamby 1997). Moreover, the plan is dynamic in nature as the social characteristics of a community, as well as communication technologies, change over time. As such, the plan must be part of a continuous auditing process that updates significant changes, such as new communication strategies. Convincing the governing body to expend funds for the evaluation process will be challenging. As Kapucu and Özerdem (2013) point out "local elected and appointed officials think of emergency management as a low priority because the public judges them on current performance and on disaster planning" (p. 44). The four reasons why the local government's risk communication plan must be evaluated include; local government's legal obligation to keep the public informed, lessons learned from other disasters, local demographic variations, and communication technology changes.

Section 252.38 of the Florida Statutes describes the County's legal responsibility to develop a Comprehensive Emergency Management Plan (CEMP) consistent with the State's CEMP. According to the statute "[s]safeguarding the life and property of its citizens is an innate responsibility of the governing body of each political subdivision of the state" (Section 253.38 FS). As it relates to risk/crisis communication, the State's CEMP mandates the county will be responsible for "ensuring the county's ability to maintain and operate a 24-h warning point with the capability of warning the public of an imminent threat or actual threat and coordinate public information activities during an emergency or disaster" (Florida Division of Emergency Management 2010, p. 15).

Garnett and Kouzmin (2007) found communication, pre and post Hurricane Katrina, to be devastating noting the breakdown of interpersonal relationships between stakeholders at the federal, state and local level. They describe a communication infrastructure hindered by the storm, causing law enforcement to use bullhorns and legwork to communicate with thousands of evacuees. Mississippi's cities and counties had to relay information to the state capital by running a vehicle back and forth during the response phase (p. 174). A risk communication plan must include exercises between first responders, private and public sectors, and nonprofits to build trust between stakeholders, and ensure effective communication strategies that reach a diverse population. Moreover, the communication plan must have an outreach component that prepares the public for emergencies.

The characteristics of a community's population change over time. Established communities together with new development and urbanized gentrification create a varied cross-section of social and ethnically diverse groups along with more modernized infrastructure. A risk communication plan must adapt to these types of changes. One of the primary components of risk communication is knowing your audience when formulating and conveying risk messages to the community (Kapucu and Özerdem 2013). The risk communicator must be keenly aware of local demographics in such areas as education level, age and ethnicity. For instance, in a community comprised mostly of senior citizens, social media technology may not be the most effective communication tool to use. The risk communicator acts as a bridge between the technical experts and the public by simplifying the message and breaking down the technical jargon into plain English (Ng and Hamby 1997).

Modern communication technologies provide an array of valuable tools (e.g. social media, internet, and cellphones, GPS, GIS) to help emergency managers effectively formulate and disseminate risk messages to a diverse public. These technologies are changing so quickly, a risk communication plan must be evaluated regularly as an opportunity to incorporate new tools into the plan and encourage bi-directional communication. Today, people are more techno savvy and want to be involved in decisions that affect their communities. A risk communication plan must include strategies for receiving information (listening to the public) in addition to transmitting risk messages. The plan must also recognize the need to train personnel on new technologies, and best practices for bi-directional communication between government and the public. As much as new technology can be seen as an invaluable tool during emergencies, if used incorrectly, the unintended consequences can be devastating. Inaccurate information can create distrust and a lack of confidence in local government's ability to handle a disaster.

8.9 Social Media and Communication

Emergency information before, during, and after a disaster is also of high importance. The public must be adequately prepared for natural disasters that are about to occur so it can lessen the negative impacts of property loss, injuries to residents and even prevent/minimize death. By providing constant information to the public (proper, accurate and well described information), emergency managers are able to better prepare the public about an impending or ongoing disaster (Kapucu et al. 2008). And in today's new age of information technology, it is necessary to have the proper social media skills and knowledge to provide information to the public. Emergency management organizations must now use social media computer based technology as part of information dissemination. By monitoring and using twitter, Facebook, MySpace, Skype, YouTube, and other social media tools, emergency management organizations can communicate with the public in emergencies and collect valuable data and information using members of the general public as sources of information on the ground. But the use of information technology social media must be done carefully because there are many falsehoods that can be portrayed in social media sites. Also, if the wrong information is given by emergency management organizations, then it must be identified and corrected quickly before more damage can be done (Latonero and Shklovski 2011).

Speaking to the skills of emergency management personnel, some organizations and agencies have requested for their employees to have social media skillsets, as mass media has been a major component of emergency management. Mitomo et al. (2013) investigated the role of mass media in post-disaster situations, specifically the Great East Japan Earthquake, and noted the impact of mass media on civic

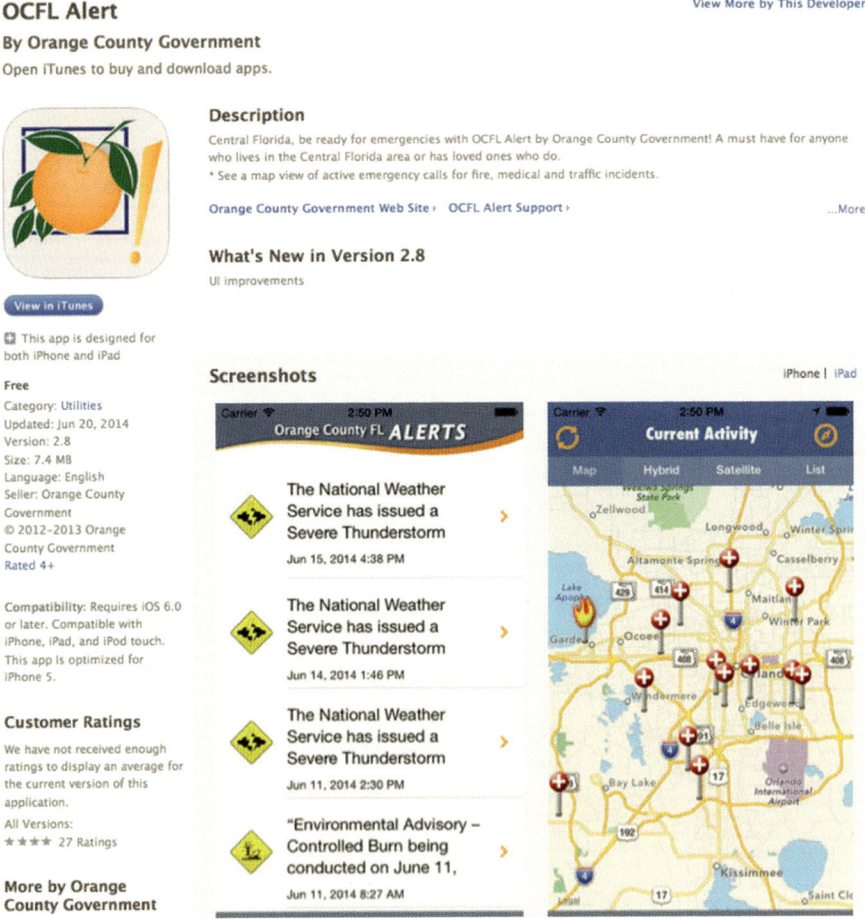

Fig. 8.3 OCFL (Orange County Florida) alert app screen capture

engagement and social capital. Through the influence of the communication mediation model, Mitomo et al. (2013) generated a theoretical model to describe the
impact of offline engagement through online participation (see Fig. 8.3). The
researchers found that community members who connect to the television programs
and web applications are more likely to participate offline and increase social capital through bonding trust and bridging networks.

Moreover, the researchers took the analysis one step further and created a model
for the perception of community members regarding disaster recovery (see Fig. 8.4).
Through testing of the model, Mitomo et al. (2013) found social capital influenced
the community's perception of recovery. Through increased participation of online
and offline engagement opportunities, social capital is strengthened and community
members are more trusting of their needs being met after a disaster occurs (Fig. 8.5).

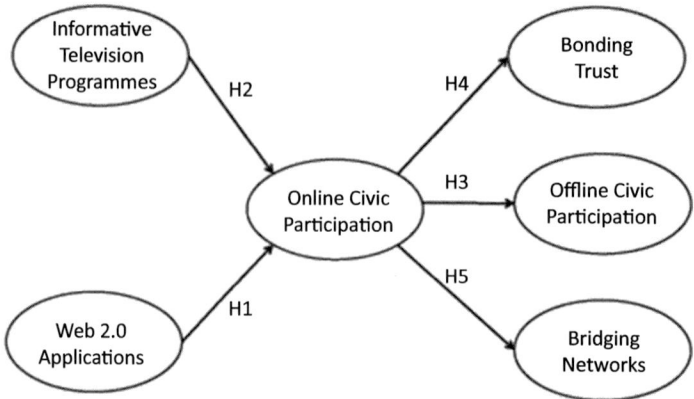

Fig. 8.4 Civic participation and social media (Mitomo et al. 2013)

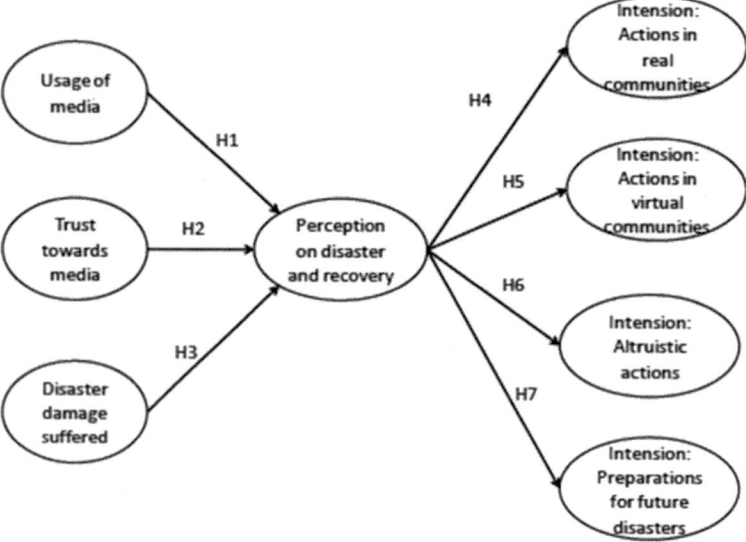

Fig. 8.5 Perception on recovery from Mitomo et al. model (2013)

Expanding the avenues for communication, many emergency management organizations and agencies entered the world of social media to engage members of the community (Appleby 2013). Some even deem social media as a way to cut out the middleman and provide a more direct connection between the response organization and community members (Flynn and Bates 2011). Through this technological avenue, the transparency of and trust for various organizations and agencies have increased.

Specific avenues of social media include Facebook, Twitter, YouTube, LinkedIn, blogs and more (Latonero and Shklovski 2011). With the increasing popularity of these portals, many organizations and agencies have incorporated social media into their public relations strategic planning and utilize the accounts to not only promote their services, but provide information regarding their mission and community events. Moreover, social media is becoming a way for communicating important information during a disaster, but these applications are helpful in the preparation and recovery phases as well (Latonero and Shklovski 2011).

An example within the state of Florida is Palm Beach County's preparedness videos targeted for their residents (Palm Beach County 2014). In response to the hurricane seasons of 2004 and 2005, county officials realized their residents lacked critical information. An approach to resolve more consequences was through YouTube videos on how to prepare for hurricane seasons. In addition, these videos connect to the county's Facebook and Twitter profiles.

One group affected by disasters, and has rarely been studied, is tourists (Sigala 2011). For states like Florida, tourists are a constant and have unique needs. For emergency management purposes, responders must be aware of tourist groups and understand their concerns in regards to communication. As tourists are implants from other areas, they are not long-term community members and may need assistance in areas like evacuations. Moreover, tourism firms are institutions experiencing crises more frequently due to globalization, urbanization, and the reliance on technology (Sigala 2011).

For example, Facebook and Twitter have been used after disasters, like Hurricane Katrina or the Great East Japan Earthquake, to help individuals locate missing family members or verify significant others were alive. YouTube has also increased in popularity as organizations can post informational and instructional videos to help prepare their communities to respond to a crisis, thereby increasing resiliency (Appleby 2013). Furthermore, social media has been used to accumulate funds for relief efforts. Some organizations have promoted monetary donations through the use of a Twitter hashtag or a text message from specific cellular providers (Flynn and Bates 2011).

With social media gaining recognition in the field of emergency management, the Federal Emergency Management Agency (FEMA) (2013) created a course in their independent study program. The objectives include: (a) explaining the importance of social media for emergency management, (b) describing major features and functions of common sites being used, (c) describing the challenges and opportunities of social media applications in relation to the five phases of emergency management, (d) discussing better practices for social media applications, and (e) building the capability of social media use and sustaining it within an emergency management organization.

This course would be incredibly useful for social media volunteers requested by Florida emergency management personnel. A drawback of social media is the perpetuation of inaccurate information and rumors. Florida is taking the issue and circumventing any significant impacts by recruiting and training volunteers to manage the social media avenues for when a disaster strikes (Kleinberg 2014). Taking the lead on this initiative is Florida State University's Center for Disaster Risk Policy. One of the center's training exercises took place at the hurricane conference where an imaginary hurricane struck a fictitious town. Conference participants utilized a designated hashtag and dispersed information to the town's citizens. This exercise helped the participants understand the benefits and challenges of social media in emergency response.

8.10   Social Media Perceptions

In this section we discuss social media platforms and their potential for sharing information for our focus group data.

There were mixed reactions to the use of social media venues to communicate disaster related information. Urban and sub-urban counties (Brevard, Orange and Seminole Counties) had a more positive outlook on utilizing social media. An example of the positive outlook was witnessed in the following exchange with focus groups participants from Brevard County:

Participant A: As technology becomes more prevalent and more available we use more of that to be about to reach people. You know, we're very big in social media here, Facebook and Twitter, you know, we're reaching out to people on a regular basis. Just this morning I tweeted about, you know, change your clocks, but don't forget to change your batteries.
Interviewer: Is social media is something that you use, has it been effective?
Participant B: It's been very effective
Participant A: I think we're probably the most active of social media of any of the counties around us.
Participant C: Absolutely, I don't think there's any county that's as active as we are in the state.

In Orange County a participant stated: "I do think that we've done a good job in the last five or six years of improving our information sharing capabilities, taking advantage of the social media now, Twitter and Facebook and our webpage". Similarly, when a asked about the place of social media in disaster communications a Seminole County participant indicated: "all those social networking sites have a viable place, um, to get the message out, because communications, you know, the more information you have, knowledge is power right? That's the old adage, so the more you get out, the more information you can provide people to have them act in their own safety, security, and wellbeing of them, themselves, and their families, the better. So it does have a viable place."

In the rural counties of Lake and Sumter County social media use was viewed with less enthusiasm, particularly the perception of a lack of capable infrastructure due to its rural nature. For instance, a participant from Lake County expressed the following concerns: "I don't see twitter and Facebook being all that utilized in

Paisley, Lake Mack, Lake Katherine, you know, um, in the eastern boarder of the county I don't see that being a, uh, resource to be able to utilize. And, I dare say that I don't think Levy County and uh, Sumter, some of those areas, are um, going to be utilizing that tool. So on the rural side, there's a challenge of access". When asked to elaborate as to why social media might not be used in the rural areas mentioned the participant explained: "Uh, it depends on the infrastructure. If the cell towers are down they're not going to be able to utilize their smartphones because their cell towers are not as robust as our public safety network. So they're not going to be able to utilize those. They won't be able to get out to the virtual world. I don't think we should not do it because of that. I just think that the infrastructure may become damaged and we won't be able to utilize it. I just don't want to – I mean it's a tool. I mean my favorite tool is slapping a big old poster board at the Jiffy Stop at the Stop and Rob (local convenience store) in the rural communities because everybody goes to the Stop and Rob."

Similarly, when asked about the usefulness of social media in Sumter County participants had the following exchange:

Participant A: Useful yes, but we don't have that kind of capacity to do it.

Participant B: And again, I'm not so sure how affective that would be in the rural area.

Participant C: There are areas can't get cell phone reception.

Participants also shared their concerns with the use of social media, particularly issues of access to the internet and whether they are effective in reaching all sectors of the public, particularly the elderly.

Evidence of these concerns is documented below:

In Orange County a participant explained: "And a lot of them are not technologically plugged in, so we have to find other means to contact them, so we do that. We use a system called Code Red where we, when we see something happening we start calling them and giving them information. Because a lot of them still rely on the telephone". Another added: "Social media does not reach everybody. Social media reaches a group of citizens. Older citizens aren't, well they're starting slowly to get into social media. (Other: Some will never do it, like my in-laws for example) Exactly, they rely on other means of communication, so, social media really is only for one sector of our citizens".

Another participant added "Yeah, it's definitely for the younger ones".

"And it might the ones that we're already reaching by about a dozen other things, because those people are also watching TV. They have everything whereas the folks, you know the older folks, probably aren't. (Others referring to seniors: they rely on the TV and newspaper; when the power goes out there's not TV they won't have information, they have no idea what's going on unless someone goes physically to their house and tells them what's going on because we had a lot of people that lost telephone, a lot of people don't have hard wire phones, or it's tied to the cable system so they lose everything)".

Similar sentiments were shared in Seminole County, for instance a participant stated: "It's going to depend on the type and extent of disaster, because, uh, twitter, and uh, Facebook, and uh, YouTube, all depend on one critical path, their back bone is the internet, and if you have a scorched earth type of disaster, where all your telephone central offices are taken out, you have no internet. Because all the internet traffic flows through Telephone Company switches. So, it's gone. Now the Sheriff's office, and I think uh, you guys do to, actually have Twitter accounts, and Facebook accounts, where we put out information for a missing child or a wanted subject that's in a particular area. Twitter, you've got 160 characters, but the Facebook, you've got a little bit more space to put in. And we're using it. Uh, a lot of people have it on their iPhones and other cell phones, which makes it now mobile. Which is good, but again, its dependent on to be mobile, the internet has to exist. The cell companies have to exist. So it is going to be something that you'll have to gauge by the event as to whether you can you is or not based on the extent of the disruption of services."

8.11 Conclusion

A community's resilience is based on a multitude of factors. Through the use of common operating pictures and information communications technology, the social capital within a community can increase and positively impact its resilience as a byproduct. Being aware of the social side to emergency management, like communication and social capital (Aldrich and Meyer 2014), impacts the community's resilience and influences the various public, private, and non-profit agencies that respond to crises. Well prepared and planned assessment, mitigation strategies, and response activities can be quickly undone if proper and effective communication is not in place. Emergency managers must prioritize risk perception and communication throughout their field, governments of all levels, and to the residents they exist to serve.

Emergency managers must utilize a variety of communication tools to ensure that proper messaging is relayed to residents. Emergency managers do not want to (and should not) be seen as "fear mongers" for a community but inspire a spirit of education, preparation, and resiliency. Communication methods can range from informational booths at community events to in-depth presentations at community meetings, and the distribution of information across social media and making resources available over the Internet. Modern technology represents one of the greatest assets to an emergency manager. Using tools such as GIS, social media, and portable devices (tablets, smartphones, etc.) enable emergency managers to assess, mitigate, prepare, and respond with increasing levels of effectiveness. Yet, there are important caveats that were addressed by our study participants such as dealing with the news media and access to the internet, among others. These perceptions message particular differences/experiences that should be noted to increase communication and disaster resilience.

References

Aldrich, D. P., & Meyer, M. A. (2014). Social capital and community resilience. *American Behavioral Scientist, 59*(2), 1–16.

Appleby, L. (2013). *Connecting the last mile: The role of communications of the Great East Japan Earthquake.* Retrieved from https://www.internews.org/sites/default/files/resources/InternewsEurope_Report_Japan_Connecting%20the%20last%20mile%20Japan_2013.pdf

Boland, S. (2008). *Guard ready for emergencies. Signal magazine.* Retrieved from http://www.afcea.org/content/?q=node/1670

Celik, S., & Corbacioglu, S. (2010). Role of information in collective action in dynamic disaster environments. *Disasters, 34*(1), 137–154.

Comfort, L. K. (2007). Crisis management in hindsight: Cognition, communication, coordination, and control. *Public Administration Review, 67*(S1), 189–197.

Comfort, L. K. (2008). *Distributed cognition: The basis for coordinated action in dynamic environments.* Paper presented at the Surviving Future Disasters, Baton Rouge, LA.

Comfort, L. K., Oh, N., Ertan, G., & Scheinert, S. (Eds.). (2010). *Designing adaptive systems for disaster mitigation and response: Role of structure.* Pittsburgh: University of Pittsburgh Press.

Department of Homeland Security. (2008). *National emergency communications plan.* Retrieved from http://www.dhs.gov/national-emergency-communications-plan-necp-goals

Division of Emergency Management. (2014a). *Comprehensive emergency management plan.* Retrieved from http://floridadisaster.org/cemp.htm

Division of Emergency Management. (2014b). *Response-logistics.* Retrieved from http://www.floridadisaster.org/Response/Logistics/Index.htm

Division of Telecommunications. (2010). *Florida emergency communications number E911 state plan.* Retrieved from http://www.dms.myflorida.com/content/download/77627/453336/Emergency_Communications_E911_State_Plan[R1].pdf

Federal Emergency Management Agency. (2013). *Social media in emergency management.* Retrieved from https://training.fema.gov/EMIWeb/IS/courseOverview.aspx?code=is-42

Federal Emergency Management Agency. (2014). *Effective communication.* Retrieved from https://training.fema.gov/EMIWeb/IS/courseOverview.aspx?code=IS-242.b

Florida Division of Emergency Management. (2010). *Comprehensive emergency management plan.* Retrieved from http://www.floridadisaster.org/documents/CEMP/2010/2010%20State%20CEMP%20Basic%20Plan.pdf

Flynn, S., & Bates, S. (2011). *Connecting America: Building resilience with social media.* Retrieved from http://cnponline.org/p/connecting-america-building-resilience-with-social-media

Garnett, J. L., & Kousmin, A. (2007). Communicating throughout Katrina: Competing and complementary conceptual lenses on crisis communication. *Public Administration Review, 67*(1s), 171–188.

Johnson, B. J., Goerdel, H. T., Lovrich, N. P., & Pierce, J. C. (2013). Social capital and emergency management planning: A test of community context effects on formal and informal collaboration. *The American Review of Public Administration*, 1–18.

Kapucu, N. (2006). Interagency communication networks during emergencies boundary spanners in multiagency coordination. *The American Review of Public Administration, 36*(2), 207–225.

Kapucu, N., & Garayev, V. (2011). Collaborative decision-making in emergency and crisis management. *International Journal of Public Administration, 34*(6), 366–375.

Kapucu, N., & Özerdem, A. (2013). *Managing emergencies and crises.* Burlington: Jones & Bartlett Learning.

Kapucu, N., Berman, E., & Wang, S. (2008). Emergency information management and public disaster preparedness: Lessons from the 2004 Florida hurricane season. *International Journal of Mass Emergencies and Disasters, 26*(3), 169–197.

Kleinberg, E. (2014). *Emergency managers stress the need for social media volunteers.* Retrieved from http://www.emergencymgmt.com/disaster/Emergency-Managers-Stress-Need-Social-Media-Volunteers.html

Latonero, M., & Shklovski, I. (2011). Emergency management, Twitter, and social media evange-
lism. *International Journal of Information Systems for Crisis Response and Management
(IJISCRAM), 3*(4), 1–16.

Mitomo, H., Otsuka, T., Jeon, S. Y., & Cheng, J. W. (2013). The role of ICT and mass media in
post-disaster restoration and recovery progress: A case of the Great East Japan Earthquake
[Conference Paper]. In *24th European Regional conference of the International
Telecommunication Society*, Florence, Italy.

Ng, K. L., & Hamby, D. M. (1997). Fundamentals for establishing a risk communication program.
Health Physics, 73(3), 473–482.

Norris, F. H., Stevens, S. P., Pfefferbaum, B., Wyche, K. F., & Pfefferbaum, R. L. (2008).
Community resilience as a metaphor, theory, set of capacities, and strategy for disaster readiness. *American Journal of Community Psychology, 41*(1–2), 127–150.

O'Brien, G., & O'Keefe, P. (2010). Resilient responses to climate change and variability: A challenge for public policy. *International Journal of Public Policy, 6*(3/4), 369–385.

Palm Beach County. (2014). *Ready south Florida: Palm Beach ready*. Retrieved from http://ready-
southflorida.org/palm-beach-county-info/

Paton, D. (2007). *Measuring and monitoring resilience in Auckland* (GNS Science report).
Lower Hutt: GNS Science.

Sigala, M. (2011). Social media and crisis management in tourism: Applications and implications
for research. *Information Technology & Tourism, 13*(4), 269–283.

Volunteer Florida. (2014). *Emergency management: About ESF-15*. Retrieved from http://www.
volunteerflorida.og/emergency-management/

Wang, X., & Kapucu, N. (2008). Public complacency under repeated emergency threats: Some
empirical evidence. *Journal of Public Administration Research and Theory, 18*(1), 57–78.

Weick, K. (1993). Collapse of sensemaking in organizations: The Mann Gulch disaster.
Administrative Science Quarterly, 38(4), 628–652.

Wolbers, J., & Boersma, K. (2013). The common operational picture as collective sensemaking.
Journal of Contingencies and Crisis Management, 21(4), 186–199.

Chapter 9
Disaster Resilience in Rural Communities

Abstract In this chapter we discuss focus groups results from rural communities and particular challenges to disaster resilience unique to rural communities; we do so utilizing the Adaptive Resilience & Community Capital framework (Fig. 6.1) introduced in Chap. 6. We discuss findings from focus group data in the rural counties of Lake, Osceola, Seminole, and Sumter Counties. Particularly, local policies and support, community capital, capital vulnerability, and private support previously identified as key to disaster resilience, adaptive resilience and adaptive learning. Furthermore, we discuss the divergent meanings of rural in Florida in comparison to other rural regions of the United States, issues of limited resources and lack of emphasis to the need of rural inland counties, self-reliance of rural communities, issues surrounding mobile homes and RV parks, and a lengthy discussion of the impact of a large suburban retirement community (The Villages) in an otherwise rural setting.

Keywords Disaster resilience • Rural communities • Rural emergency management • Adaptive resilience • Community capital • The Villages • Florida

9.1 Rural Emergency Disaster Management

In a majority of rural communities in Florida, disaster response operations are often handed through volunteers (Kapucu et al. 2013). Although weather patterns make many regions of Florida vulnerable to hurricanes, other hazards such as droughts, wildfires, tornados, are routinely present in rural counties. These risks are compounded by agricultural operations, which place high demands on the response by local officials. Rural emergency disaster managers tend to face more difficulties in their respond to hurricanes due to financial constraints, lack of training and equipment (Janssen 2006), and a lack of fiscal resources from an un-diversified economic base. Furthermore, state and federal grants based on population size as a formula may be biased against rural communities, thus impacting their efforts to develop and implement hazard mitigation practices (Caruson and MacManus 2008). The resource constraints of communities are more pronounced when considering infrastructure vulnerabilities. Antiquated public infrastructure is a source of

© Springer International Publishing Switzerland 2015   121
F.I. Rivera, N. Kapucu, *Disaster Vulnerability, Hazards and Resilience*,
Environmental Hazards, DOI 10.1007/978-3-319-16453-3_9

vulnerability in smaller, more rural communities because these areas lack the resources to rebuild or undertake recovery efforts resulting in a large financial burden (Chang and Shinozuka 2004).

Older housing structures, located in areas prone to disasters and are non-conforming to building codes, are susceptible to long term damage. Mobile homes, for instance, are being constructed with stronger frames and greater resistance for wind gusts, yet evacuation orders will still affect these individuals as no one can truly predict the capabilities of the storm itself against the structures (Kusenbach and Christmann 2013). Therefore, vulnerabilities for mobile home communities become considerably dependent upon the capabilities of these structures to withstand disaster events. For emergency managers, there is an increased risk of financial impact in regards to rebuilding and finding homes for displaced individuals.

In addition, the demographic makeup of rural communities brings about certain challenges with the inclusion of poor and older populations. Rural areas with low-cost homes, such as mobile homes, include individuals with lower incomes, lower levels of education, and livelihoods dependent on resource-based occupations, such as farming (Prelog and Miller 2013). When disasters strike, the impact can be seen in the recovery efforts in terms of sprawl and job loss. If the disaster zone takes away economic opportunities, then the individuals deemed vulnerable are placed in a more precarious situation to make ends meet (Kapucu et al. 2013). Furthermore, older populations have increasing health needs and a reduced earning capacity. Combined with issues of space and isolation, poorer and older residents play a big factor in the capacity of rural communities to prepare and response to disasters (Waugh 2013). Moreover, it becomes the responsibility of emergency managers to anticipate their communities' needs, as citizens may not make explicit demands (Henstra 2010).

9.2 Lessons from Florida Communities

9.2.1 Rural Florida

Previous research on the demographics characteristics of rural communities in the US indicate they are likely be older (Waugh 2013), poorer (Fothergill and Peek 2004) and lack access to essential services (Prelog and Miller 2013) in comparison to their urban counterparts. One of the emerging themes from the focus group data was the perception of rural being different in Florida than in other areas in the U.S. (e.g. rural Midwest). A respondent from the Sumter County group stated:

> Well one of the things to think about when you say rural, or use the term rural around here versus the mid-west, rural in Florida is not the way it is out in the heartland of America. You still have a mix of people that grew up native to the area, then you have a lot of people that have come down to retire and have bought five or ten acres of land. And uh, in Sumter County, we've got a really, what I call a bizarre mix of people in here. We've got the Villages (retirement community) which is 80 % of our population, that's a large retirement

community, a suburb; and then you have the rest of the county which is that mix that I'm talking about. Plus, we've got a lot of transient people in recreational vehicles that come and stay for a couple weeks at time or whatever. Um, so, our challenges are different than what you'd say in rural America. You know, that's a different setting, it's a different culture, it's a different environment. We've got a lot of people that are not from here, we've got some people that are from here, and we've got a lot of people passing' through, and uh, it's a different mix of education uh, cultural backgrounds and what not.

Yet, the aforementioned issues of having an older population base and a perceived lack of resources were still present in the discussion of rural communities. For instance, a participant stated: "Uh, it's an interesting mix and when you start looking at it and talking about what all the issues are, and we talk about special needs, um, a lot of people come down here in their late 60s, early 70s, maybe even early 80s to retire. Um, typically the male will die off first and the wife will be left alone and the kids are still up north, and so, we find that's our typical special needs type client that we have to take care of because nobody else can take care of them. They don't have any real caregivers in the area. So, they may need to be in the shelter, they may not, it's just uh, just kind of a challenge that we have to go through and uh, that's, that could be rural Sumter, that could also be suburban Sumter."

Furthermore, "So when we're talking about rural Florida, the situation is the same in the other Florida counties. Lake County saw the same thing. Clermont was a very agricultural citrus related community back in the 80s, the freezes came, the land was developed and now you've got this similar situation. All the counties in Central Florida, and of course South Florida, were like this at one point. The term rural does not necessarily, in our minds, we don't see it the same way as other people when they use the term rural; a horse and a trailer, you know that kind of thing. This is a kind of a unique mix of different situations that, uh, Old Florida is transitioning into a new Florida, and uh, our demographics are such that, um, most of new Florida is over that age of 55."

Another participant stated: "And um, in South County, from an infrastructure perspective, we don't have the economy of scale because the folks are all spread out, so you lose the economy of scale. To fix a road, or fix a light, or fix something in the city you fix it for a hundred people. Fix it out here, you fix it for one or two. Plus, our roads in South County start out at a lower level than the roads elsewhere. So you've got (background noise, inaudible) and then there are all the non-county 9–11 roads, little private entrances, but you know, 16 people live on this little piece of dirt, and it starts low, and then when a storm comes it's gone. So, those kinds of things become real problematic for public works to try to work with emergency services and try to bring them back. It takes a lot of effort, and I don't want to say a little gain for a lot of effort, but it's a lot of work for a small number of people. That's a problem I've seen in the rural part of this county." Indeed, previous work by Waugh (2013) suggests the impact of space, which renders the provision of services more expensive in rural communities. Compounded with economic struggles responsible for the reduction of resources limits the capacity of emergency management. As shared by a participant in Seminole county: "I think it's just, probably, there are only so many people to go around in the EOC (Emergency Operations Center), as far as funded positions. Um, I think we can do great in the field, and we have people out there, but we don't necessarily have all of the resources in there (EOC) as a county government just because of government cut backs."

9.3 Adaptive Resilience & Community Capital Framework Elements

9.3.1 Local Policies & Support

The inland location of the rural counties in our study region was perceived as a particular element providing some insights on a few of the issues particular to rural communities, as witnessed in this conversation in the Lake County focus group:

Interviewer: "You mentioned that in terms of the political structure, you know, because you're inland, you're not Miami or Tallahassee, or Orlando or any other major area, does that affect you also? In terms of the political risk?"

Participant: "Oh yeah. I tell people all of the time, we are an inland county. You will see resources go to the crust and we will strive to get the best resources that we can but we will get what's left. Um, you know? And while I appreciated all of the help on Ground Hog Day, for some folks that gives them a false set of what we might get during a hurricane. Because they saw all of this help rolling in and everything like that, but, however, during a hurricane we are not going to be the impacted area and we are not going to get those resources. And there is an acknowledgement of that. And I think our elected leaders understand that. I think we've had good luck with that in the past, but you're right. In terms of the hurricane center, you know, it doesn't just affect Lake County. It's going to affect other places harder."

Interviewer: "You mentioned that the things about Disney is that they take care of their people, but there is the possibility that people from the coast might go into the inland counties and –"

Participant: "They steal our gas! We lose our gas on 27 and 441 and 50. They take our gas. I mean it's an economic impact for us because our gas goes away because they're traveling through and they're going north. So they take our gas. And uh, that's going to be, because the gas is going to go to the crust, and I've still got to figure out how to get gas to keep us going. And I think that's another thing that is not recognized in emergency management. Particularly in the resilience and the recovery phase of it is, we have to make, we make decision and we have to influence things to make sure that our community survives and thrives economically. Because it's not just about getting somebody's house repaired. It's about making sure that they have a place to go to work, and that the power is back at their work. So, um, we talk about, okay this is critical facilities and this infrastructure and all that other stuff. But I mean, I've got 27, 441, and 50, that I've got to make sure power is going to because that's my economy. And um, then we fan out. Luckily all my critical infrastructure sits on those three roads, so it works out. But, you know, um, it's, in rural communities it can be tough to balance that. Um, you know, we just have, we have one particular area that is a challenge. I guess when you get down into Green Swap there can be some issues, but that, that's very sparsely populated. Uh, and I got one Stop and Rob down there where I know that I need to get resources to and they will feed that whole community. But man, you get stuff down in there (referring to the "Stop and Rob") and that whole community just blossoms. And that's the other thing, just knowing your community. Knowing which ones, like Lake Mack. Lake Mack is great, they've got the Stop and Rob right there at the intersection of the highways. The problem is that it's got a small little, uh, parking lot, so getting things in and out of there is a challenge. But you can get information to them."

Another participant added: "And that's key, what he's talking about, you know? The POD – the point of distribution, because you'll have people coming in and they don't know where to go, and if you can make it a one stop shop. You know? You can have the medical team there, diapers, supplies medicine, fuel and food. And then there's neighbors helping neighbors and that begins the recovery rather than saying "there's no help here" even if they have to walk to where they can get help, it's available. And then that can be where the point gets communicated back to the EOC there's injured or nothing left here and we need assistance. Those PODs are just huge assets."

9.3.2 Pizza Crust

An interesting perspective from one of the study participants in relation to the perception of the system as unfavorable to those in inland communities was the analogy of the pizza crust. In detail:

> I think one of the challenges is because Florida, the analogy that I use, Florida is like a pizza crust. Because the crust is where the majority is where the population is and that's where the biggest impact to a land falling hurricane will be. And, the faith-based (organizations) go to those densely populated, high impact areas and, and, here when the storm comes in it's a little bit of rain and some wind. But, on the coastal regions it's devastation and carnage, so that's where, I know I've had experience where there has been a tremendous influx of that. So a lot of that is self-deployed and you arrive to assist and find out that (unclear). You know and we don't get a lot of assistance inland, but you also have to look at what's going to be the biggest benefit to your population and usually that occurs on the pizza crust of Florida and a lot of people are resilient enough themselves.

When asked about any other differences between those in coastal communities and those in rural communities, the participant added: "Well the biggest thing is that the people in the pizza crust move to get out of the way of the storm inland. So not only do you have residents but you have a transient influx of squatters trying to wait to get back to their coastal communities, so that's one of the dangers that I don't think Florida has been impacted by yet but Disney just, the area around Orlando just fills up from people running' from the coastal area. And, you know they run from water and hide from wind and that's usually where they go is inland. And even so in the coastal community's public safety will put assets inland and try to make their way back. So, I hate to say it but the center of the state is neglected because the storm starts to de-intensify, you know, as it comes in. And Florida is only mainly thinking about hurricanes you know, it's the sudden tsunami or the sudden impact, it's the other all hazards events that Florida will be going, you know, "what was that?" That's why I think Florida in the sense that you get a three-day window of it's coming, where other areas don't. You get 20 minutes and sirens and then it's over."

9.3.3 Isolation

Another theme identified was the perception isolation of some in rural communities. As one responded stated: "I was just thinking, that the areas that you all work in too get the south part of the county, being so rural and even knowing that there is a problem going on with some of the people down there that must be so difficult. When you start getting down to Kenansville and Yeehaw and some of the ranches. There's a lot of trailers out there that you wouldn't know about unless you stumbled across them."

9.3.4 Community Capital

One important element to disaster resilience is the community capital an area processes. Although our focus group participants identified several constraints, they also perceived a strong sense of community capital, particularly the perception of the self-sufficient nature of rural residents: "going back to the rural area, a lot of those people are pretty resilient up there. They're pretty resilient because they choose to live out in that rural area and they know everything is farther away from them and, not that there aren't some needs out there, because by all means there are." Another participant in Osceola County added: "The good news, I think, is down in the south part (rural area). They're very self-sufficient and they're very neighborly and they will help each other out and they have the equipment to do so. And I think that's what they have going for them. They're very self-sufficient. And sometimes I think we take that for granted."

The view of rural communities as self-reliant is commendable, but, as Waugh (2013) warns, this view might not be as accurate as imagined:

> Self-reliance is often assumed to be strength of rural communities in the United States. That is the idyllic picture of the American heartland, but it is not the reality today. Undoubtedly there are still some communities characterized by strong social networks and a willingness to help neighbors in need. But, communities have changed as family farms have been displaced by agri-businesses, often owned and run by international conglomerates. Farms may

be owned and run by part-time farmers who hold full-time jobs in local malls or factories. The point is simply that the image of rural America is less romantic than it was decades ago and rural communities are less resilient than they used to be. Even food and water for local consumption may be scarce commodities in regions that produce food. Clearly, it is also difficult to generalize about rural communities because all are not agricultural and all are not distant from major urban areas (294).

Another identified theme was the strong sense of community and the perceived different impact a disaster might have. This sentiment was captured in the following quote from a Sumter County focus group participant:

> I think some of the unique challenges that rural communities have is, uh, in an urban area, your neighborhood next door can be wiped out for whatever reason, and a lot of times that doesn't affect you. But in rural communities, if your neighbor gets hit, you know your neighbor, or you're related to your neighbor and, and, it's a greater impact to the community and so one persons impact is, is spread throughout the community a lot quicker. It's harder to bounce back. So I think those are some particular challenges to the community and because it is a close community, a rural community, if you lose a family of three, in a particular area, that's going to have a lasting impact on that community. You know, if you have large segments where people are significantly hurt, or killed, it's going to have a very lasting impact mentally, for the mental health of the community.

9.3.5   Capital Vulnerability

Some of the themes identified incorporated examples of capital vulnerability (Fig. 6.1), are the narratives in relation to mobile homes and recreational vehicle (RV) parks. For instance, a participant from Osceola County shared the following: "The mobile homes are a different thing, and that has presented a number of issues, because it's the shelter of choice for uh, those with low and moderate incomes, and yet it's the most vulnerable. Uh, we've looked at reclassifying areas that, so they don't get rebuilt with mobile homes that in itself are a battle. Sometimes with the property owners, sometimes with their tenants. And so then you create a situation where the tenants and the property owner have different objectives. Mobile homes in and of themselves I think really are problematic when it comes to major events."

Similar issues were raised when discussing RV parks, as expressed by a Sumter County participant: "RV parks, we've got lots of 'em and they're growing. And most of them, I hope this is correct, most of them aren't here during the hurricane season. There are some here, but even when it's not hurricane season, strong winds or tornadoes would just wreak havoc. And they're not prepared for it and I don't think we're prepared for it." Another participant added: "The thing about an RV park is that it's not always the same people. I mean, I have friend that comes down to Florida and he camps in a different one just about every time. He has a fifth wheel and picks whatever one he wants to park in, and it's not always in Sumter County. Uh, so, people in the RV park that are truly vacationers or transients for the summer, or winter, may not be the same ones that were there last year. So they don't really know the county that well, they don't know what's available or the lines of communication. They might not even know where the EOC is or any of that, they don't know that it exists."

9.3.6 Private Support and Other Elements

The influx of older populations moving into suburban developments within rural areas is a particular issue of concern for emergency managers. One of the biggest developments of this kind is The Villages, located in Sumter County. The Villages was developed in the mid-1980s as an older adult community (Bartling 2006). As of 2014, The Villages had 107,056 residents making it the fastest growing metropolitan area in the United States (Kunerth 2014). The following exchange exhibits some of the particularities of this residential developmental touching upon role issues of private support, community capital, and capital vulnerability previously discussed as elements for adaptive disaster resilience.

9.3.7 Real Rural vs The Villages

When asked to describe the disaster resilience of the community, focus group participants in Sumter County shared their perception of the differences between those outside of the Villages and those inside of the Villages. For instance, a respondent stated: "I think that we were talking about people that live on one or two acres of land and stuff like that. They're kind of independent. And I'm talking about the non-villages. So we really have two communities here, we have, or two demographics. You have the Villages, which is sort of like a suburban area, and they're not so much independent, although as a group they are. And then you've got the rest of the county where people are independent. They take care of themselves, they take care of their property, they take care of their family, their neighbors, uh, I think that's an asset for them."

Another participant added: "And a lot of them actually have machinery and equipment that plays right into that. I mean, it's not uncommon that if they have a large brush fire or something of that nature, it's not uncommon for them to actually mitigate those and control it and deal with it, and help each other out. It's not uncommon to have a neighbor god to somebody else's house and say "hey let me help you with that". Uh, which is a huge asset, uh especially in a county that is rural and doesn't have, um, you know in terms of numbers, doesn't have a large population to go out and be able to respond to those types of incidents."

Interestingly, there was also a perception for those outside of the Villages who do not have the expectations for disaster emergency management governmental services verses those who reside in the Villages. As one respondent mentioned: "And they also don't have the expectation, for the uh, let me phrase this correctly.

The people south of the Villages don't have a high expectation of the government in terms of services. The people in the Villages, they have more expectations because the Village is a governmental entity mixed in with a corporation, under a county government. So they have a high expectation that their roads are maintained and all the others things that go along with it. And as soon as there's a hiccup we know about it. It takes a while for the folks south of that to let us know there's an issue; sometimes they fix it themselves, sometimes they deal with it. I call it the electric can opener test. If the power goes out and you can open your canned goods because you're dependent an electric can opener, that's the kind of, you know, that we're gonna run in to. Some people know how to, you know, use a manual can opener, and some people don't have a manual can opener. Uh, some people call it the Waffle House test too. I've heard that term before." Another respondent added: "What's interesting is when the Ground Hog Day tornadoes came through, and they hit the majority of the Villages when they went by, um, the roads were cleared so quickly and everything was handled so fast, but a lot of that was done by (Other: the corporation), well the corporation, but it was done by people who live, or work in the community and were the contractors in the community. The folks who actually resided there were not the ones that were actively involved in it."

The role of private support brings about an interesting challenge for disaster emergency managers as, on the one hand, there is access to quick and reliable private support, but mostly unable to County residents not living in the Villages; however, on the other side, it creates the potential of unrealistic expectations which might not be meet accordingly with the reduction of funding for emergency preparedness and other elements previously mentioned. Various focus participants stated these issues. One reported: "That's a unique aspect of the Villages, is, uh, it's basically the largest sub-division in the United States. It has, will soon have, 10 CDD's, community development districts, and each one is a quasi-governmental thing. This Villages Inc., the corporation that's building this and then turning it over to the CDD is still a live and well, building 200 homes a month. They are economically motivated. When something does go bad, they jump in with all feet flying' and they fix things. Now I wasn't here the last time, but I understand that they just turned their contractor loose and they fix it and they clean it up because they don't want the residents complaining' and they don't want the residents to tell prospective buyers that it's going to lay around for months." Another stated: "I think the residents themselves sheltered everybody within six hours. There was no one displaced that didn't have a roof over their head the same evening." Another added: "So that's sort of the unique thing up there that isn't replicated elsewhere (in the county)."

9.3.8   But How Will It Be When the Developers Leave?

This was a particular issue of concern express by the focus group participants, as detailed in the following exchange: "I mean, you have a large fire up there, a large brush fire, they have tankers, they have water, all of those amazing pieces of heavy machinery that they could bring in literally in a matter of minutes. I mean, how many bulldozers and dump trucks are running' around up there? But once they build out, all of that responsibility, where is it going to lay? Where is that resource going to be? And that's a huge, huge risk down the road." Another respondent stated: "So it's an asset now, but there's a level of expectation that's being built on it and it'll go away eventually." The concerns were also addressed by another participant: "And that dynamic will probably change within 5 years or so, when they've built out. Of course nobody knows what in their mind, but the, uh, the Villages Inc., will step aside and the people actually, the individuals districts manage it. They each have their own budget now, and they decide how they're going to spend it, but they do it based on direction and input from this corporation, which as you mentioned, it's a marketing thing. If there's something wrong in the Villages, the corporation is going to run an fix it because they don't want it to get on the TV, they don't want it to be seen that there are three houses within the Villages where the roof has been blown off and they want it fixed right away. That won't exist and then it'll be back to the people that are around this table, uh, to fix it, as any normal government would."

When asking about other differences between the Villages and the rest of the county, respondents shared the following:

> Well I think there is a similarity, and we talked about how people in the rural end of Sumter County are resilient because they're self-reliant, but a lot of people who live in the Villages are people who come from retired military, retired law enforcement, fire, um, these are retired people who come from something where they, they were involved in these kinds of responses in their home town; maybe Illinois, Massachusetts, New York, wherever it is, they come down here and yea they're retired, but they didn't leave that life behind. So, I think there is a similarity between that guy that lives on the five acres down in Coleman whose next neighbor is a half a mile away, and the fireman who lives in Duval in the Villages, uh, it just needs a little chemistry maybe to kick start it to get it going. But they're pretty much the same I think. The only thing is that the guy in the Villages doesn't have the resources that the guy down there in Coleman has. He doesn't have a tractor sitting' in the back, he doesn't have the things that he might be able to use to do on his own. So he has to rely on government agencies to come in, even though he might be very capable of doing some of that stuff, he just doesn't have it anymore. I'm a Villager, I used to repair all my own cars. I can't even change my oil now. I don't have the stuff to do it. So that's pretty much where it is. People, uh, can do all those things, but they've chosen to let go of the things that allow them to do it.

Another individual added:

> The other thing about the Villages, and I live there too, is um, that they have a lot of educational opportunities, and I don't mean just simply schooling, but they have fairs about storms, and they have programs on the radio and TV, which they have their own stations, on how to prepare for a hurricane, what to have and everything. They have their own CERT, Citizen Emergency Response Team. They have their own radio club, they have groups of people that are doing things. We have neighborhood watch going on and everything. So on the one hand, they move in from Massachusetts, on the other hand when they get here the indoctrination starts. You're a Villager, what are you going to do? Are you going to be in CERT or are you going to join the radio club, or what are you going to do? And remember we have hurricanes, and they go the hurricane expo and everybody tells them what to do, and they have little bags that you can pick up and put all the stuff you need in it and everything. So there is a very pro-active educational sort of civilization going on in the Villages.

And there are over 90,000 people up there, and I mention that because, uh, there was a comment about how soon will they stop building. Well the goal has always been to get to 100,000, so we're pretty darn close and if the goal doesn't move. You know the goal post keeps moving. If the goal doesn't move, within a year or two, we'll be at our goal. So yeah, you might find that the administration turns things over and starts shutting down the actual building part of it, but we'll have to wait and see. But it could be sooner than we think.

The exchange continued with the following response from another focus group participant:

The good news is that the CDD (Community Development District) structure will remain, it legally has to. There will be 10 community development districts, and then there is one call the VCCCD, the Villages Central Community Development District, that will sort of, it's the straw that stirs the drink, and they have their own public safety entity up there, they've got their own fire department that works with Sumter County Fire and Rescue. So they've got infrastructure that even after the DRI stops building, they're a leg above some of the other communities. But they don't have everything; they won't have that Villages Inc. contractor. And we know 'em, they can respond in a blink of an eye with massive amount of people and equipment, and it's really cool. But that will go away in a few years. The other thing is, and we've sort of touched on it. The uh, how the Villages respond and how the residents responds, sort of depends on what time of the year it is. Between January and May the Villages, I don't know if it doubles, but the population increases hugely with part-time residents – the snowbirds are here. They are not quite as well trained, they do, some of them get involved, and some of them don't. June they're gone, and it get's back down to a steady state from June til I don't know, October or November and these are the people who they know the community, they know what to do. They've probably taken more training, and it's a smaller number, probably a little bit more trained, but if something bad happened in February, my goodness, we've got a lot of people that just came down for the week, or for the month and suddenly we got a disaster. My suspicion is that it would not bode well.

Other perceived notifiable differences and similarities, between the Villages and the rest of the rural county residents, include issues with sense of community, resources, experience, and preparedness. Beginning with a sense of community and resources, a participant shared:

I think though, even when the contractors aren't involved anymore, one of the cool things about the Villages is that the folks there do have a desire to help; a neighborly desire to help each other. They have a pretty significant AED program that is set up throughout the community where they'll buy and AED, a community will and mount it on a house, and they have a pager system and stuff, so maybe they don't have the physical resources, to be able to bring heavy machinery, so they may be reliant on your first response and your government agencies to kind of get them through that hump so to speak. But after that, they have the financial resources to bring to bear, that probably the rural portion of the county that doesn't have as much demographics, doesn't have enough population to support the financial needs, they have 90,000 people in one area that have financial resources that can help them get back on track. Where if you remove the income for the other part of the county, they either of some sort of retirement investment or sort of financial solid foundation, you know they're living, either working on their retirement or living whole heartedly on retirement, the rest of the county may not be in that type of situation.

Another participant added: "Plus I would wager that there are more people in the Villages that have insurance than in the rest of the county."

Yet, there were some perceived notions of caution fueled by the makeup of the Villages residents. As one respondent stated:

> And not to be hard on the Villages but if there was a significant disaster up there, that really would be a problem because they, you know, they're together – it's easy to affect a lot of them with the same thing. You know, one tornado drops down there and you might affect 1,000 homes. If it drops down somewhere else here you might kill some cows, and you know might knock over a tree, and maybe even ruin somebody's barn. But if you drop one in the Villages, you've got a lot of people who are affected. The other thing to is that you've got a lot of medical facilities up there, and uh, uh, getting those things back up and running the way they were before, and they are relied upon by the 90-something thousand people that live there. All of the sudden you don't have those doctors, um, now these people are looking' for them. If that happened here, they probably would go to some other, Brooksville, or whatever, up that way, they can't get to it by the golf cart. By the way, it's a golf cart community. It's going to be a major problem, and being able to bring that infrastructure back to support that is going to be a lot harder than bringing it back to another part of the county. I don't know if we're really a rural county or not. We're more rural-retirement if anything, because we have a lot of RV parks, we have a lot of people who are here that come, I don't know what our population, we talk about the Villages, but the whole of Sumter County is affected by that. Right here on the main street, you've got this new RV park that's gone up, down the end here, I mean how many of our population goes away after Easter and doesn't come back until October or November?

When ask how prepared the Villages are to the rest of the county, respondents had divergent views. For instance, one person stated: "If you ask a lot of them they'll say yes. But, uh, I don't necessarily agree with that. Uh, there's, uh, the strength in that community is that they have good educational campaigns; they have good community emergency response teams. They're aware of the fact that they're vulnerable to hurricanes and stuff like that, but are they more prepared individually? I don't know." Yet, another participant disagreed: "As a community I would agree, as a community the Villages is prepared. You know, we talked about the CERT, public safety, and the community watch that drives around; as a community yes, as individuals no. Because of the fact, as I said, they've transitioned from one type of life, to a retirement life, and they've left a lot of their infrastructure that they used to carry around with them behind. They don't have it anymore. So, they just don't have the ability to do something the things they could have done back in Massachusetts or Illinois, they don't have it here. As individuals, I would say the rest of Sumter County is much better prepared to help themselves. In the villages, they have to look to the community because they just don't have the stuff in their own homes to do it."

There was also a perceived view of when the Villages benefit from having strong community bonds and resources, the residents still do not have sufficient experience to deal with a significant disaster event. This is confirmed by one participant:

> It's funny because I was just at a, um, educational campaign back about two weeks ago, at Walmart here in Bushnell, and uh, getting people to converse with us about hurricane preparedness and what not, a lot of times what we heard was "Oh, uh, I've lived here all my life, I'm prepared, I know where it floods, I know what happens in hurricanes," and blah, blah, blah, but if we were to do the same thing up in the Villages area, um, there's a lot of people, that come from the north that have never gone through a hurricane, you know, and they can't associate what the wind and water can do. Um, and we see that quite a bit with the new homeowners. They're more; they're more interested in what can happen than the

people who have been around for a while. They don't seem to be as interested in it because they've lived here for 20 years or so and they've never seen a hurricane or tornado, I know it was bad in 2007, but I can't convince the rest of the county because it was only that segment of the Villages that got hit with the tornado. Hurricane Frances and Jeanne, as much damage as they did here were not hurricanes when they hit, they were tropical systems. Um, a straight on, right off the Gulf Coast hurricane, no one has that experience here. So, a lot of people think they've gone through big hurricanes and tornados but really they haven't, and they're going to get surprised when we have that big catastrophic event.

9.4 Conclusion

In this chapter, we identified some of the issues perceived by rural counties focus group participants in relation to the elements to disaster resilience previously discussed in Chap. 6. In particular, we discussed the divergent meanings of rural in Florida in comparison to other rural regions of the United States, issues of limited resources and lack of emphasis to the need of rural inland counties, self-reliance of rural communities, issues surrounding mobile homes and RV parks, and a lengthy discussion of the impact of a large suburban retirement community (The Villages) in an otherwise rural setting. Each issue was discussed along with their impact to the elements of the Adaptive Resilience and Community Capital framework. The findings from the focus groups provide valuable lessons of the specific issues affecting disaster resilience in rural communities, which if left unattended have the potential to create other vulnerabilities and hazards.

References

Bartling, H. (2006). Tourism as everyday life: An inquiry into The Villages, Florida. *Tourism Geographies, 8*(4), 380–402.

Caruson, K., & MacManus, S. A. (2008). Disaster vulnerabilities: How strong a push toward regionalism and intergovernmental cooperation? *The American Review of Public Administration, 38*(3), 286–306.

Chang, S. E., & Shinozuka, M. (2004). Measuring improvements in the disaster resilience of communities. *Earthquake Spectra, 20*(3), 739–755.

Fothergill, A., & Peek, L. A. (2004). Poverty and disasters in the United States: A review of recent sociological findings. *Natural Hazards, 32*(1), 89–110.

Henstra, D. (2010). Evaluating local government emergency management programs: What framework should public managers adopt? *Public Administration Review, 70*(2), 236–246.

Janssen, D. (2006). Disaster planning in rural America. *Public Manager, 35*(3), 40–43.

Kapucu, N., Hawkins, C., & Rivera, F. I. (2013). Disaster preparedness and resilience for rural communities. *Risk, Hazards, and Crisis in Public Policy, 4*(4), 215–233.

Kunerth, J. (2014, March 27). The Villages: Retirement community is nation's fastest-growing metro area. *Orlando Sentinel*. http://articles.orlandosentinel.com/2014-03-27/news/os-villages-fastest-growing-census-20140325_1_the-villages-elaine-dreidame-mayor-ed-wolf

Kusenbach, M., & Christmann, G. (2013). Chapter 4: Understanding hurricane vulnerability. In N. Kapucu, C. Hawkins, & F. Rivera (Eds.), *Disaster resiliency: Interdisciplinary perspectives* (pp. 61–83). New York: Routledge.

Prelog, A. J., & Miller, L. (2013). Perceptions of disaster risk and vulnerability in rural Texas. *Journal of Rural Social Sciences, 28*(3), 1–31.

Waugh, W. L. (2013). Management capacity and rural community resilience. In N. Kapucu, C. Hawkins, & F. Rivera (Eds.), *Disaster resiliency: Interdisciplinary perspectives* (pp. 291–307). New York: Routledge.

Chapter 10
Farmworkers and Resilience

Abstract In this chapter we analyzed data from a focus group with farmworkers in Central Florida to investigate disaster resilience in rural America. We identified three major themes within our coding scheme: past disaster experiences, self-organizing collective action, and challenges to self-organizing collective action and resilience. The results indicate disaster experiences can serve as a pathway to disaster resilience. In addition, we discuss significant challenges and barriers continue to be present including, language/communication issues, anti-immigrant sentiment, poor relations with the police and farm owners that serve as constant reminders of the vulnerabilities and challenges migrant farmworkers continue to face, not only in Florida, but also throughout the United States (A previous version of this chapter has been accepted for publication at the *International Journal of Mass Emergencies and Disasters*).

Keywords Farmworkers • Disaster resilience • Collective action • Language • Communication • Immigrants • Risk perceptions • News media • Culture • Florida

The examination of a community's vulnerability is critical for disaster resilience (see Chap. 4) as it provides a sense of the community's ability to cope, prepare and recover from impacts of natural hazards (Donner and Rodríguez 2008; Waugh 1994). Many rural regions have a robust agricultural economic base threatened by natural disasters. Often the work available in rural labor markets is low wage and low-skill (Lichter 2012). This type of work has attracted recent Hispanic immigrants who lack education, and who have considerable language and cultural barriers (Donato et al. 2007; Farmer and Moon 2009; Kochhar et al. 2005). The influx of new immigrant populations, an aging population, and a constant threat of disasters add unique challenges to building disaster resiliency in rural communities.

Much of Florida remains rural. Of the 406 cities, 89 % (363) have populations less than 50,000 and 80 % have less than 25,000. Among the 67 counties, 46 % (31) have a population of less than 50,000. In Central Florida there are approximately 87 rural communities (U.S. Census Bureau 2000). These areas, like other rural communities in the U.S., experience a lack of fiscal resources from an un-diversified economic base and have an antiquated public infrastructure. These challenges are a source of vulnerability in smaller and more rural communities because these areas

© Springer International Publishing Switzerland 2015
F.I. Rivera, N. Kapucu, *Disaster Vulnerability, Hazards and Resilience*,
Environmental Hazards, DOI 10.1007/978-3-319-16453-3_10

may lack the resources to rebuild or undertake recovery efforts because of a potentially large financial burden (Chang and Shinozuka 2004). Older housing structures, located in areas prone to disasters, are non-conforming to building codes and susceptible to long-term damage. Also, approximately two million residents, constituting 12.5 % of Florida's population, live in mobile homes (Caruson and MacManus 2008; Schreiber 2005).

10.1 Disaster Resiliency and Farmworkers

The bulk of research on migrant farmworkers discussed their vulnerabilities including limited access to health services (Carrion et al. 2011; Castañeda et al. 2010), a lack of social integration (Bail et al. 2012; Lichter 2012), substandard housing conditions (Ziebarth 2006), and lack of trust of government (Chavez et al. 2006) among others. Research on the experience of this group in disaster situations find instances of marginalization including discouragement for seeking shelter by the presence of border patrol agents at shelters and disaster checkpoints, (false) accusations of stealing relief resources, and deportation of those in shelters that did not have appropriate documentation (Mathew and Kelly 2008; Nuñez-Alvarez et al. 2007). The experience of the Watsonville farmworkers in the 1989 Loma Prieta earthquake demonstrated governmental and other aid resources were not culturally appropriate to meet the needs of this population (Tierney 2007). In all, few studies have examined disaster resilience as it relates to this particular population of rural America.

10.2 Rural Communities, Farmworkers, and Strategies in Building Resilience

The data for this chapter come from a series of semi-structured focus groups conducted between November 2011 and March 2012, as part of a larger project analyzing rural disaster resiliency in Central Florida (see Sect. 1.2.2 for a detailed description). We based our results from data from the Volusia County focus group conducted in the rural town of Pierson, Florida. Volusia County is located in the east section of Central Florida and covers an area of 1,207 square miles. There are around 1,114 farms covering more than 229,000 acres and produce a variety of agricultural products including fruit, vegetables, honey, cattle hay, sod, fish, timber and plants (Volusia.org 2013). The economic activity generated from Volusia's agriculture and natural resources industries are responsible for an estimated $781 million annual economic impact and represent three percent of the county's economic activity (Volusia.org 2013). The Pierson area main products are ferns used in flower arrangements. Laborers in this area work in ferneries and some wok in citrus groves in other areas of Central Florida (Farmworker Association of Florida 2013). In 2010, Volusia

	Florida	Volusia	Pierson
Population	18,801,310	494,593	1,736
White	75 %	82.5 %	57.5 %
Black	16 %	10.5 %	4.8 %
Hispanic/Latino	22.5 %	11.2 %	54.1 %
65 years and over	17.3 %	21.1 %	10.2 %

Table 10.1 Selected demographic information for the State of Florida, Volusia County, and the Town of Pierson

County accounted for 2.6 % of the total population of Florida, making it the 11th most populous county in the state (Florida Legislature Office of Economic and Demographic Research 2012) (Table 10.1).

10.3 Key Findings

To examine rural disaster resiliency, the results from the Volusia county focus group, taken in the town of Pierson in which farmworkers were well represented, was analyzed. The examination and coding was derived from a concept driven perspective know as framework analysis (Gibbs 2007; Ritchie and Lewis 2003) and incorporated a list of thematic ideas drawn from the disaster resiliency literature. Afterwards, members of the research team read the transcripts and assigned sentences to different thematic domains based on the interview script. At the conclusion of this process, a meeting took place to review the themes until consensus was reached. Research participants were identified as respondents with no identifiable traits outside of gender. Based on the interview script, the concept-driven themes, and the research team consensus, participant quotes were divided into the following themes: past disaster experiences, self-organizing collective action, and challenges to self-organizing collective action and resilience.

Ten respondents participated in the focus group, including representatives from the Volusia County Office of Emergency Management (VCOEM) and the Farmworker Association of Florida. Eight of the ten participants were local farmworkers invited to participate in the focus groups by a representative of the Farmworker Association of Florida Pierson area office, the remainder participants were from the VCOEM.

10.3.1 Past Disaster Experiences

Participants in the focus group related their experiences with disasters, primarily the 2004 hurricanes Charley, Frances, and Jeanne, and shared their experiences with the tornadoes, which affected the area at the end of 2006 and the beginning of 2007. One participant stated, "…because there had never been hurricanes…and they came hard one after another…some three weeks in a row…but, in reality, we weren't prepared…we as organizations…like this one but in reality we weren't prepared…". He went on and shared about the lack of help after the hurricanes stroke the region, "but in reality the help came late because, FEMA came after eight or ten days…and there was no light for about five days…and the people in Seville (nearby town) didn't have anything to eat…there was no gasoline…and we suffered."

Other experiences touched upon the health challenges aggravated after the disaster experience. For instance, a young woman described her experience as follows: "There are a lot of kids here with asthma…most are born with asthma and when it is a time like this is worse…As we had…when there was hurricane, the roof fell on top of us…it was on a day that we weren't at work…when we came back to the house we found a terrible mess, everything thrown off the bed, everything was a mess in the bathroom too, a lot of things covered in black (dirty) and after that it was worse because one of our children has bad asthma and with this was getting worse."

The contributors also related confusion on where to go for shelter. A woman recalled her experience: "Well at first we were very scared, you know? …because it was really strong……And we all went to the school…and this (muffled)…because they said that they were going to help make a big room…right?…that was strong (i.e. safe)…because if a hurricane or something came…because they won't let us get into the school again…because this time the roof of the school was lifted."

Overall, the farmworkers in the focus group had a torrid disaster experience leading them to organize in the hopes of having a better response in the event of a future disaster. As one participant stated about the hurricane experience: "I think…it was a good experience…for the community." Indeed, these experiences started the process of self-organizing collective action, which we discuss below.

10.4   Self-Organizing Collective Action

Collective action refers to joint action in the pursuit of a common objective (McAdam and Snow 1997). Self-organizing processes for collective action recognizes the limitations of traditional methods for coordination, like through a third party. Feiock (2009) refers to this as a constructed network. This type of mechanism is designed or coordinated by third parties such as higher-level government to structure relationships across actors. A higher-level authority provides funds and incentives for actors to participate in collaborative relations and designates a lead organization with responsibility for developing, managing and coordinating processes.

In comparison, self-organizing systems emphasize the networks of interactions among actors who foster norms of trust and reciprocity. A network of exchange relations among individuals can emerge unplanned. One underpinning of self-organizing systems, advanced by scholars, is collective-action theory linked to social capital. For instance, Ahn and Ostrom (2008) view social capital as "an attribute of individuals and of their relationships that enhance their ability to solve collective-action problems" (p. 71). Putnam (1995) suggests social capital is "social organization such as networks, norms, and social trust that facilitate coordination and cooperation for mutual benefit" (p. 67) and can lead to more efficiency and social benefits (Putnam 2000; Rohe 2004). The emergency management profession and scholarly work on resiliency, more specifically, has demonstrated a shift toward examining the complex social relations contributing to reducing vulnerability of major disaster events. From this perspective, bonds are one of the primary drivers of a sense of community, place attachment, and citizen participation enabling individuals to be emotionally connected, voice their concerns, mitigate potential limitations of population diversity, and fairly distribute the roles and responsibilities (Aldrich 2010; Norris et al. 2008). Nonetheless, more emphasis needs to be placed on the human element of disaster resiliency and how this contributes to a community (Kusenbach and Christmann 2013).

Emphasis on the role of interpersonal relations in strengthening the capacity of a community to withstand major disaster events has been echoed at the highest levels of government. For example, according to the Federal Emergency Management Agency (2011), collective management needs to include government and non-governmental organizations working closely with individuals, families and communities, as they are the nation's most important assets as first responders during a disaster. Collective collaborative approach emphasizes the principles of individual empowerment, partnership, and inclusiveness to locally led recovery organizations and processes. This approach is also reflected in the Whole Community perspective, guided by FEMA, for the preparation and response to disasters. In some cases, these principles have been "formalized" through interorganizational networks, such as Community Organizations Active in Disasters (COADs), which coordinate with local emergency managers to engage in pre-disaster planning (Gazley 2013).

After the disaster experiences, farmworkers decided to organize and reach out to other groups to be ready in the event of another natural disaster. Several rural farmworker groups, like the Farmworkers Association, the Alianza de Mujeres Activas, Alliance of Active Women, (AMA), and Hispanos Unidos (Hispanics United), got together at the instance of the Florida Catholic conference. In more detail, a representative from VCOEM stated:

> But from that experience is when we started to unite factors…we didn't know, you know, that this organization didn't know what AMA was up to, so there was a lot of disconnect and what not, and that experience kinda brought us to the table and brought us together… and I will continue on the history on the Florida Catholic conference and the person is in charge of disaster management with the, entire state, state of Florida…all the Catholics, he brought a lot of groups together here in this building in a meeting and that's when we started to realize…wow, you did this, you did that…why don't we do it together?

In conjunction with the Volusia County Office of Emergency Management, the Farmworker Association of Florida, along with other members of the community, created the "Grupo Comunitario de Respuesta a Desastres," or Disaster Response Community Group. As one of the participants stated: "…we've created this group. We have certified people, many of us, like some 15 people in the Community Emergency Response Teams (CERT) group…you know…we made this little back room for disasters…it's the only thing we have in this rural area…to help every-one…in this area." Furthermore, one of the VCOEM participants stated:

> He squeezed six years of hard work into two sentences and you know what's happening as a result of that 2004 hurricane season is that we have about 15, actually 21 citizens, emergency response teams members trained…Alejandro is one of them, Maria is one of them, I'm one of them, I'm one of the trainers, Marcos received his training…others that are not in this group have received formal training thanks to county resources…we got a lot to tell…but then also as a result of starting to work together and resources coming together and seeing networks and what not…this room was built…this room as a disaster community center..

She went on and indicated:

> We've been meeting regularly since 2006 and invited to the Univision (Spanish Channel) health fair several years ago and in one of those health fairs we had a meeting with the com-munications office and they said how can we improve things? And what they got in the works was a contract that was eventually signed with the county of Volusia and Entravision, which is the parent company for Univision and their radio services, that they would carry the disaster news that came out of Volusia county and others. It's been my observation since then that Univision really has step up to the plate and you do get more. I used to say at the meetings, unless Univision tells it in the middle of the soap operas this community isn't going to catch the news.

The creation of the group, collaboration with other agencies/groups, and dis-semination of emergency warning created a path to disaster preparedness not pres-ent before. As one member relayed, the farmworker community feels more prepared in case of a disaster: "Now we feel a little more [prepared]…we know where…we know that we are going to look for help here or if we don't find the help at least we have adequate knowledge."

On the surface, it seems networking and collaboration has created a path to disas-ter resiliency, which was absent before. Nonetheless, there are ongoing barriers and challenges focus group participants shared with us and will be discussed in the next section.

10.5 Challenges to Self-Organizing Collective Action and Resilience

There were several challenges noted by the focus group participants, particularly issues dealing with language, anti-immigrant sentiment, relations with the police and farm owners, the reliance on volunteers from the VCOEM, and the lack of work after a disaster.

10.5.1 Language

Even though the farmworkers were able to organize and had Univision helped with emergency warnings, the focus group participants felt the need for more disaster related information in Spanish. As one individual recalled: "There is no communication because of language. It is an obstacle… the people who don't speak English do not communicate with the people who are not Hispanic…except those Hispanics that speak English speak with them…" Another participant noted:

> We have a lot of information in Spanish…because there are many people, right?…but still we lack a little more that…because we are a diminished group and there are many people around us that sometimes cannot be reached…and so many people are lacking knowledge…and then I believe that the news, the means of communication, I believe they should help more to cover and so that the people are more informed…because, sometimes the media, if you watch something other than Univision does not pass on the information regarding what is happening… it's what we see more between the Hispanic news or newspapers…I think that a lot of people…and I think there needs to be more done that more information from the different organizations, from FEMA, the Red Cross, What is it that they do? How could they help more quickly…I think that here we are a little lazy regarding progress, you know?…and well, more resources to extend themselves, to give more information, you know?

10.5.2   Anti-immigrant Sentiment

One particular concern for this group of farmworkers is their perception of a growing anti-immigrant sentiment, which creates an atmosphere of fear and distrust of disaster emergency organizations. For instance a participant commented:

> I think that we are missing that trust and that there should be reform, I believe that the people would be more calm, attend where they can and…now there are many limitations… because the anti-immigration laws they want to bring in from different states and that Florida wants to impose, we are splitting legislations to try to keep them from happening in Florida…I think that this is the fear of the majority of the people…And this happened with a lot of people during the hurricanes, they didn't get help out of fear…that is the government, it's going to fight immigration, not all the people, but many people who are distrustful…I think that there needs to be some confidence created for these people.

10.5.3   Relations with Law Enforcement

The relations with police are a constant struggle for this group of farmworkers. Several respondents shared their experiences and views on the relations with the police. A male participant stated, "here the police, rather than keeping the people confident/trustful, they keep them fearful…" A woman added: "…I am scared to drive, because I could get stopped by the police, and I had my license before and then when it expired, I was without a license for one month because of my expired license, the police stopped me and took my license and now they've stopped me three times and told me I can't drive for five years and for this reason I don't attend the meetings because I am afraid to drive…" Another female respondent further described her experience, "I am alone with my children and I am also scared to walk around, because the police stop me and…it's the same thing, because someone who doesn't have a license…they know that someone doesn't have a license or they go to jail or pay a lot of money, but the job that one has is not enough, nor the police… because the fines are some $300, it's a difficult situation…".

10.5.4   Relations with Farm Owners

Relations with farm owners are equally distrustful. Review, for instance, this exchange:

Interviewer: And tell me about the relation with the bosses, What about an emergency situation? Do they make sure you are ok?

(Laughter) Various voices: They make sure that they themselves are ok…

Interviewer: So it has to be the community that…

Woman: Instead of increase the salary? They lower it…they lose ferns so they lower the salary to gain a little more for themselves, to recuperate what they lost.

Man: They don't lose anything, they have insurances/security.

Woman: Friday we had to have a meeting with the boss and he said that they had lowered the salary because they had lost a lot of ferns and that they had to decrease salaries to ensure that they could keep employing workers…(Various voices)

Interviewer: And do you think in any situation the bosses would help in a disaster situation?

(Various voices): No…

Man: No because when the Red Cross comes, although there was a disaster, many bosses tell them (to the workers) they don't need to eat and the bosses said that the workers don't need food, that everything was ok…

Interviewer: And why?

(Various voices): It's a mystery.

10.5.5   Volunteers in Disaster Response

Farmworkers in the focus groups felt the need for an established person to be part of the county office of emergency management. In the works of a group participant, "…but I think the county needs an established person (inaudible), that is bilingual, so there is more trust with the people…" A volunteer with the VCOEM agreed and declared, "Thank you for pointing out that what I've done is as a volunteer for the last 6 years, and I thank you…so the county should really appoint somebody who is bilingual that can recognize, and part of their job would be to approach this community…" While another participant shared his views, "I think that we lack resources so that people can be more constant (confident), because sometimes we do a lot of different work and we don't dedicate ourselves to what is really a disaster. We give information, all that we can, but…as far as volunteers go I think, some people that have time could give out this information for pay, and could dedicate themselves to this and only this…".

10.5.6   Lack of Work

An additional concern for farmworkers is the constant uncertainty of available work, particularly after a disaster situation, and was shared by a focus group participants: "On the other hand, for the farmworkers I think that when there is a disaster…the rural workers, this is when they suffer most, because in reality the insurance companies pay for our work…but the farmworker remains without work and sometimes there is no help for them…in reality this is when they suffer a lot…there is no work, the leaf (fern) that is very small dries up and afterward they are working 2 or 3 days, looking for selectos (leaves they can harvest), in a drying furrow of (inaudible)…this is the only crisis that if there had been here…also the freezes."

Another stated, "Things haven't changed, I think they've gotten worse…because there's less work…(work has gone down) and the bosses go bankrupt…and they don't want to make a new company… we have to look other places for bosses that offer jobs, because here they don't want to offer jobs, they only offer jobs to their own people…".

10.6   Conclusion

In this chapter, we analyzed data from a focus group with farmworkers in Central Florida to investigate disaster resilience in rural America. We identified three major themes within our coding scheme: past disaster experiences, self-organizing collective action, and challenges to self-organizing collective action and resilience. The results indicate disaster experiences can serve as a pathway to disaster resilience. In a sense, communities can utilize these experiences as an opportunity not only to "bounce back" to pre-disaster functioning, but also to actually leap forward (Rivera and Settembrino 2013). The findings from the self-organizing collective action theme indicate this community of farmworkers was able to utilize their previous disaster experience to create a network of groups to collective take measures to actively prepared and plan for a disaster event. The collaboration of the farmworkers groups with the County Office of Emergency Management suggests the challenges of rural areas require local leadership. It also implies rural development is in need of a multifaceted approach incorporating social and effective economic development strategies. Communities need multijurisdictional and multi-organizational network mechanisms (Bradshaw 1993; Kapucu et al. 2013). Kapucu and Garayev (2011) argue collaborative emergency focuses on the networked coordination to tackle disaster events. They also emphasize a decentralized and flexible structure that incorporates relevant administrative and service delivery adjustments. Collaborative emergency management has become an inevitable, let alone indispensable, tool to deal with complex extreme events over the last years (Waugh and Streib 2006).

Participation by residents in developing resilient communities is critical. The analysis of this dilemma is informed by the broader body of work in the fields of community development, planning, public administration, and urban studies. This research base argues expanding the scope of citizen participation produces more responsive local policies. One of the three models of citizen input, discussed by Godschalk et al. (2003), is the advisory model. This approach relies on citizen input through public hearings and committees. The public hearing is designed to afford citizens the formal opportunity to give comments on proposed plans, ordinances and projects to local elected officials. A second model is based on participatory planning theory emphasizing collaborative planning. In this approach, citizens and stakeholders are given significant roles and degrees of power in creating and selecting alternatives (Forester 1999; Innes 1996; Wondolleck and Jaffee 2000). This model focuses on building implementation capacity by decentralizing power and enabling information sharing among stakeholders. The third model, discussed by Godschalk et al. (2003), is built around conflict management and resolving the disputes that arise when participation brings stakeholder groups into opposition (Godschalk et al. 1994; Susskind et al. 1999). Methods of consensus building and dispute resolution emphasize facilitated negotiation processes and mediation. As researchers note, the advisory, collaborative, and conflict management theories tend to intertwine.

However, heterogeneous populations increase the difficulty in coordination actions and tasks. The characteristics of the local population may influence the extent residents participate in government centered programs and activities. For example, studies on migrant workers and immigrant populations suggest there can mistrust of government officials. As a result, many of the most vulnerable populations are not fully engaged in process and are marginalized in decision making.

The complex social system shapes community development efforts in general and disaster resiliency efforts more specifically. Bradshaw (2000), for example, suggests the increased scale of relationships, the differentiation among the components, and intensifies levels of interdependence between system units. Bradshaw and Blakely's (1979) earlier work suggested rural community development involved three phases. The first phase incorporates technical assistance by experts to farmers, industry or communities. The second phase consists of organizing community members into groups through which technical information and strategies can be implemented more efficiently. The third phase emphasized a shift to networks among organizations for sharing resources, collaborating on projects, meeting multiple community needs and building capacity in the interrelations among agencies to solve problems. This third phase requires collective action and is difficult because residency suggests communities consider the multifunctional aspects of and multiple needs rather than one issue. A more recent study of the third phase suggests, "community development networks in poor communities have moved from being strange to being the favored institutional fabric out of which effective community development initiatives must be constructed" (Bradshaw 2000, p. 144).

Even though this particular community was able to self-organize, significant challenges and barriers continue to be present. The sub-themes discovered (i.e. language, anti-immigrant sentiment, relations with the police and farm owners, the reliance on volunteers, VCOEM, and the lack of work) are constant reminders of the vulnerabilities and challenges migrant farmworkers continue to face, not only in Florida, but also throughout the United States. Lack of work is of particularly importance to communities who predominately farm, fish, log, or mine, as they are typical examples of resource dependent communities (Bailey and Pomeroy 1996; Marshall et al. 2007). When the nature of the relationship between users and a resource changes, it can compromise the resiliency of social systems by altering the ability of the user groups to retain essential community functions and undermine prosperity (Adger 2000; Burdge and Vanclay 1996; Farmer and Albrecht 1998; Marshall et al. 2007). Furthermore, there is a need to integrate race, culture, and language into emergency preparedness risk communication programs and policies in racially and ethnically diverse communities (Andrulis et al. 2007).

Lack of trust is another vital process to acknowledge as previous research suggests trust-based partnership and collaboration is important for effective emergency response and recovery operations (Kapucu 2006). Risk perception and communication are interrelated in the field of emergency management. The perception of risk by individuals at any time is related to how effective and efficient information was communicated as well as how promptly and correctly individuals utilize that information. How individuals perceive the risk and prepare themselves for both natural and manmade disasters is critical for disaster preparedness as the process of hinges on the public of informing those who are impacted by these disasters (Kapucu and Özerdem 2013).

Understanding the different methods of communication, such as active listening, helps first responders and the public to accurately communicate risk. Knowing the culture, language, sources available for the community is a vital part of ensuring information is not only given correctly, but also received in the manner of importance for which it is striving to achieve (FEMA 2011). Effective risk communication can help to reduce the complacency individuals gain as a result of their current perception of the risk (Kapucu et al. 2008). Complacency is defined as "a sense of confidence or self-satisfaction that is created by ignoring danger" (Kapucu and Özerdem 2013, p. 183). When risk is communicated accurately, continually, and effectively at all stages of disaster management individuals are able to identify the risk, which leads to public readiness.

Using Hurricane Katrina as a case, Garnett and Kousmin (2007) identify four lenses for communication to be implemented during a crisis. The lenses are interpersonal influence, media relations, technology showcase, and inter-organizational networking. The lens utilizes leaders and key emergency personnel in working together and building relationships before, during, and after a crisis has occurred. This assists the community in identifying the areas of risk and resources needed prior to an event occurring. This will also allow emergency management leaders with the opportunity to know their community and vice versa. The media relations lens includes those who report the news and the representatives of the organizations involved with the current event. The media can assist with informing the community of an event, showing the current danger or risk involved and the recovery stage of the event. The media, if not informed in advance of the background of an organization or meeting, can also give the wrong perception during a crisis. The third lens is technology showcase. There are various forms of technology, which can help or hinder communication action implementation. It is still said the best form of communication is face-to-face, but technology, such as the national weather system, emergency alert system, public service announcements, radio, text messaging, email, etc., has the capability to reach a large number in a short amount of time. The fourth lens is inter-organizational networking and includes individuals in public, private, and the nonprofit sector working together. Communication implementation is only done well in this lens if all organizations trust one another. The sharing of resources and intelligence allows the reaping of large benefits from this lens during a crisis, but the lack of trust and increase competition and rivalry decreases the value of this lens for crisis communication.

Identifying and addressing risk perception of individuals in communities is as important as selecting the correct communication channel as leaders must gain trust of the people before a crisis to increase the probability of the community heeding to their warnings. Understanding different methods, and knowing both when and how to use them correctly, will give the message sender confidence their note is being received by emergency managers and community leaders and incorporated into a plan to prepare the community for a crisis.

Finally, the findings from this chapter suggest self-organizing collective action can be effective in creating disaster resilience, even in socially vulnerable populations. Nonetheless, the results also indicate the recurrence of barriers is a constant reminder of the goal in creating truly disaster resilient communities not being reached if these conditions are not lessened or eradicated.

References

Adger, W. N. (2000). Social and ecological resilience: Are they related? *Progress in Human Geography, 24*(3), 347–364.

Ahn, T. K., & Ostrom, E. (2008). Social capital and collective action. In D. Catiglione, J. van Deth, & G. Wolleb (Eds.), *The handbook of social capital* (pp. 70–100). Oxford: Oxford University Press.

Aldrich, D. P. (2010, June). Fixing recovery: Social capital in post-crisis resilience. *Journal of Homeland Security, 6*, 1–10.

Andrulis, D. P., Siddiqui, N. J., & Gantner, J. L. (2007). Preparing racially and ethnically diverse communities for public health emergencies. *Health Affairs, 26*(5), 1269–1279.

Bail, K. M., Foster, J., Dalmida, S. G., Kelly, U., Howett, M., Ferranti, E. P., & Wold, J. (2012). The impact of invisibility on the health of migrant farmworkers in the southeastern United States: A case study from Georgia. *Nursing Research and Practice, 40*(1), 119–129.

Bailey, C., & Pomeroy, C. (1996). Resource dependency and development options in coastal southeast Asia. *Society and Natural Resources, 9*(2), 191–199.

Bradshaw, T. K. (1993). Multicommunity networks: A rural transition. *Annals of the America. Academy of Political and Social Science, 529*, 164–175.

Bradshaw, T. K. (2000). Complex community development projects: Collaboration, comprehensive programs, and community coalitions in complex society. *Community Development Journal, 35*(2), 133–145.

Bradshaw, T. K., & Blakely, E. J. (1979). *Rural communities in advanced industrial society: Development and developers*. New York: Praeger.

Burdge, R. J., & Vanclay, F. (1996). Social impact assessment: A contribution to the state of the art series. *Impact Assessment, 14*(1), 59–86.

Carrion, I. V., Castañeda, H., Martinez-Tyson, D., & Kline, N. (2011). Barriers impeding access to primary oral health care among farmworker families in Central Florida. *Social Work in Health Care, 50*(10), 828–844.

Caruson, K., & MacManus, S. A. (2008). Disaster vulnerabilities: How strong a push toward regionalism and intergovernmental cooperation? *The American Review of Public Administration, 38*(3), 286–306.

Castañeda, H., Carrion, I. V., Kline, N., & Tyson, D. M. (2010). False hope: Effects of social class and health policy on oral health inequalities for migrant farmworker families. *Social Science & Medicine, 71*(11), 2028–2037.

Chang, S. E., & Shinozuka, M. (2004). Measuring improvements in the disaster resilience of communities. *Earthquake Spectra, 20*(3), 739–755.

Chavez, M. L., Wampler, B., & Burkhart, R. E. (2006). Left out: Trust and social capital among migrant seasonal farmworkers. *Social Science Quarterly, 87*(5), 1012–1029.

Donato, K. M., Tolbert, C. M., II, Nucci, A., & Kawano, Y. (2007). Recent immigrant settlement in the nonmetropolitan United States: Evidence from internal census data. *Rural Sociology, 72*, 537–559.

Donner, W., & Rodríguez, H. (2008). Population composition, migration and inequality: The influence of demographic changes on disaster risk and vulnerability. *Social Forces, 87*(2), 1089–1114.

Farmer, F. L., & Albrecht, S. L. (1998). The biophysical environment and human health: Toward understanding the reciprocal effects. *Society & Natural Resources, 11*(8), 707–717.

Farmer, F. L., & Moon, Z. K. (2009). An empirical examination of characteristics of Mexican migrants to metropolitan and nonmetropolitan areas of the United States. *Rural Sociology, 74*, 220–240.

Farmworker Association of Florida. (2013). *Pierson area office*. Retrieved on January 16, 2013, at http://floridafarmworkers.org/index.php/pierson-office

Feiock, R. C. (2009). Metropolitan governance and institutional collective action. *Urban Affairs Review, 44*(3), 356–377.

FEMA. (2011). *IS-242a effective communication* at http://www.training.fema.gov/EMIWeb/IS/is242.asp

Florida Legislature Office of Economic and Demographic Research. (2012). *2010 census county profiles*. Retrieved from http://edr.state.fl.us/Content/area-profiles/2010-census-county/index.cfm

Forester, J. (1999). *The deliberative practitioner: Encouraging participatory planning processes*. Cambridge, MA: MIT Press.

Garnett, J. L., & Kousmin, A. (2007). Communicating through Katrina: Competing and complementary conceptual lenses on crisis communication. *Public Administration Review, 67*(1), 171–188.

Gazley, B. (2013). Building collaborative capacity for disaster resiliency. In N. Kapucu, C. Hawkins, & F. Rivera (Eds.), *Disaster resiliency: Interdisciplinary perspective* (pp. 84–98). New York: Routledge.

Gibbs, G. (2007). *Analyzing qualitative data*. Thousand Oaks: Sage.

Godschalk, D. R., Parham, D., Porter, D., Potapchuk, W., & Schukraft, S. (1994). *Pulling together: A planning and development consensus-building manual*. Washington, DC: Urban Land Institute.

Godschalk, D. R., Brody, S., & Burby, R. (2003). Public participation in natural hazard mitigation policy formation: Challenges for comprehensive planning. *Journal of Environmental Planning and Management, 46*(5), 733–754.

Innes, J. (1996). Planning through consensus building: A new view of the comprehensive planning ideal. *Journal of the American Planning Association, 62*(4), 460–472.

Kapucu, N. (2006). Interagency communication networks during emergencies: Boundary spanners in multi-agency coordination. *The American Review of Public Administration, 36*(2), 207–225.

Kapucu, N., & Garayev, V. (2011). Collaborative decision-making in emergency and crisis management. *International Journal of Public Administration, 34*(6), 366–375.

Kapucu, N., & Özerdem, A. (2013). *Managing emergencies and crises*. Boston: Jones & Bartlett Publishers.

Kapucu, N., Berman, E., & Wang, S. (2008). Emergency information management and public disaster preparedness: Lessons from the 2004 Florida hurricane season. *International Journal of Mass Emergencies and Disasters, 26*(3), 169–197.

Kapucu, N., Hawkins, C. V., & Rivera, F. I. (2013). Emerging research in disaster resiliency and sustainability: Implications for policy and practice. In N. Kapucu, C. Hawkins, & F. Rivera (Eds.), *Disaster resiliency: Interdisciplinary perspective* (pp. 355–358). New York: Routledge.

Kochhar, R., Suro, R., & Tafoya, S. (2005). *The new Latino south: The context and consequences of rapid population growth*. Washington, DC: Pew Hispanic Center.

Kusenbach, M., & Christmann, G. (2013). Understanding hurricane vulnerability: Lessons from mobile home communities. In N. Kapucu, C. Hawkins, & F. I. Rivera (Eds.), *Disaster resiliency: Interdisciplinary perspectives*. New York: Routledge.

Lichter, D. T. (2012). Immigration and the new racial diversity in rural America. *Rural Sociology, 77*, 3–35.

Marshall, N. A., Fenton, D., Marshall, P. A., & Sutton, S. G. (2007). How resource dependency can influence social resiliency within a primary resource industry. *Rural Sociology, 72*(3), 359–390.

Mathew, A. B., & Kelly, K. (2008). *Disaster preparedness in urban immigrant communities: Lessons learned from recent catastrophic events and their relevance to Latino and Asian communities in Southern California*. Los Angeles: Tomás Rivera Policy Institute.

McAdam, D., & Snow, D. A. (1997). Introduction: Social movements: Conceptual and theoretical issues. In D. McAdam, & D. A. Snow (Eds.), *Social movements: Readings on their emergence, mobilization, and dynamics* (pp. xvii–xxvi). Los Angeles: Roxbury.

Norris, F. H., Stevens, S. P., Pfefferbaum, B., Wyche, K. F., & Pfefferbaum, R. L. (2008). Community resilience as a metaphor, theory, set of capacities, and strategy for disaster readiness. *American Journal of Community Psychology, 41*, 127–150.

Nuñez-Alvarez, A., Martinez, K. M., Ramos, A., & Gastelum, F. (2007). *San Diego firestorm 2007 report: Fire impact on farm workers and migrant communities in North County*. San Marcos: National Latino Research Center, California State University.

Putnam, R. (1995). Bowling alone: America's declining social capital. *Journal of Democracy, 6*, 65–78.

Putnam, R. (2000). *Bowling alone: The collapse and revival of American community*. New York: Simon and Schuster.

Ritchie, J., & Lewis, J. (2003). *Qualitative research practice: A guide for social science students and researchers*. Thousand Oaks: Sage.

Rivera, F. I., & Settembrino, M. (2013). Sociological insights on the role of social capital in disaster resilience. In N. Kapucu, C. Hawkins, & F. Rivera (Eds.), *Disaster resiliency: Interdisciplinary perspective* (pp. 48–60). New York: Routledge.

Rohe, W. M. (2004). Building social capital through community development. *Journal of the American Planning Association, 70*, 158–164.

Schreiber, S. (2005). *Mitigating the effects of hurricanes in Florida: The challenges of upgrading mobile home parks*. Tampa: University of South Florida, School of Architecture and Community Design.

Susskind, L., McKearnan, S., & Larner, J. T. (1999). *The consensus building handbook*. Thousand Oaks: Sage.

Tierney, K. J. (2007). From the margins to the mainstream? Disaster research at the crossroads. *Annual Review of Sociology, 33*, 503–525.

United States Census Bureau. (2000). *Census 2000 gateway*. Retrieved from http://www.census.
    gov/main/www/cen2000.html
Volusia.org. (2013). *The University of Florida/Volusia County Extension*. Retrieved from http://
    www.volusia.org/services/community-services/extension/
Waugh, W. L., Jr. (1994). Regionalizing emergency management: Counties as state and local gov-
    ernment. *Public Administration Review, 54*, 253–258.
Waugh, W. L., & Streib, G. (2006). Collaboration and leadership for effective emergency manage-
    ment. *Public Administration Review, 66*(s1), 131–140.
Wondolleck, J., & Jaffee, S. (2000). *Making collaboration work: Lessons from innovation in natu-
    ral resource management*. Washington, DC: Island Press.
Ziebarth, A. (2006). Housing seasonal workers for the Minnesota processed vegetable industry.
    Rural Sociology, 71, 335–357.

Chapter 11
Tourism and Resilience

Abstract In this chapter with provide an overview of the tourism and disaster literature, follow by a discussion of the emergency management with regards to tourism in Florida. We conclude the chapter with an examination of focus group data in relation to tourism and disaster resilience.

Keywords Tourism • Disaster resilience • Transient populations • Emergency management • Evacuation • Strategic planning • Risk perception • Florida

Tourism is an important economic force for the State of Florida, particularly in Central Florida. According to Visit Florida Research (2014), in 2013 there was $76.1 billion in tourism spending and there were more than one million tourism related jobs. Tourism in Florida is the leading driver for the state's economy. Florida drew approximately 93.7 million tourists in 2013 (Visit Florida Research 2014) of which 78.8 million were domestic, 11.2 million were from overseas (primarily from the United Kingdom, Brazil, Venezuela, and Argentina), and 3.7 million were from Canada. For the field of emergency management, the demographics of an area can generate unique needs should a disaster strike. With a focus of restoring services as quickly as possible after an event, emergency managers must be cognizant of the tourist population to mitigate any potential impacts.

As the number and frequency of disasters increase, along with their impact, it is important to understand the nature and scope of disasters when proactively developing strategies for mitigating, preparing, responding and recovering from such incidents (Irvine and Anderson 2006; Ritchie et al. 2003). The effect of disasters on a destination's image can influence the demand, positively or negatively (Avraham 2006). These strategies should integrate ways to reduce or limit the impact of disasters on the unpredictable tourism industry. This is especially critical for Florida as tourism is a major source of income for many state's citizens and is geographically exposed to a multitude of hazards (see Chaps. 3 and 4).

Our examination of focus group data related to the tourism industry found that perceptions to be mostly positive; however, there were some lingering issues and experiences we which discuss in detail after providing an overview of the tourism and disaster related literature.

© Springer International Publishing Switzerland 2015   151
F.I. Rivera, N. Kapucu, *Disaster Vulnerability, Hazards and Resilience*,
Environmental Hazards, DOI 10.1007/978-3-319-16453-3_11

11.1   Disasters and Tourism

Tourism, or the activities of persons traveling to and staying in places outside of their usual environment for not more than one consecutive year for leisure, business, and other purposes (Fayos-Sola 1996), has received significant attention in the disaster and emergency management literature. The information is critical as the tourism industry continues to grow and is considered a major element of today's global economy. The emphasis of the research has been dedicated to specific threats to the industry, such as transportation emergencies (either weather related or human-made), lodging fires, climatological and geographical hazards (e.g. hurricanes, snowstorms, and floods), and interruptions in business operations (e.g. labor disputes, technological failures, etc.).

In all, tourists are viewed as a vulnerable population to account for in all phases of the disaster emergency process: mitigation, preparedness, response, and recovery. A significant volume of the literature emphasized the response and recovery of the tourism industry to adverse disaster situations. Early in the 1990s, Thomas Drabek brought attention to the issues of disaster response and evacuation in the tourist industry (Drabek 1991, 1994a, b, 1995a, b, 1996). Within his work, Drabek (1995b) posited the stress-strain framework theory related to emergency management and tourism focused on organizational analysis. This framework utilizes five key interpretations for administrators and incorporates the areas of behavior, judgment, and planning. A crucial discovery was the necessity for the disaster management plans and strategies to be created by those who will implement them. Furthering the implications, Drabek (1995b) narrowed down to four aspects when creating and implementing management policies:

1. The tourist industry represents a vulnerability of catastrophic potential, but the risk is not fully recognized by those within it.
2. Community partnerships comprised of local emergency managers and tourist industry representatives should be initiated to stimulate greater awareness of the current vulnerability and to encourage implementation of preparedness plans.
3. The leadership within tourist industry trade associations and professional organizations should initiate more activities to increase an awareness of and support for disaster evacuation planning.
4. Educational initiatives should be implemented to insure that university curricula in tourism, travel, and hotel administration include more emphasis on disaster management, including mitigation, preparedness, response and recovery (p. 14–16).

Bridging off of Drabek, recent research supplements the implications through focus on sub-issues for tourism and disaster including: risk perception (Floyd et al. 2004; Kozak et al. 2007), safety (Pizam et al. 1997), and travel destination choices (Law 2006). Moreover, there is a call to increase research in the area of tourism disaster management focusing on increasing reduction and readiness strategies and initiatives (Pennington-Gray et al. 2011; Ritchie 2008).

11.2   Tourism and Emergency Management in Florida

Tourism disaster managers, emergency disaster management has put some emphasis on tourism-related disaster planning. For instance, the Federal Emergency Management Agency's (FEMA) Higher Education Project added tourism into a course curriculum guide for collegiate instructors to prepare managers. Narrowing down to Florida, the University of Florida (UF) (2012) created the Tourism Crisis Management Institute in 2007 whose mission is:

> To address the critical need to prepare tourism industry professionals for crises, the Tourism Crisis Management Institute developed a comprehensive Online Tourism Crisis Management Certificate for Destination Management Organizations (DMOs), Attractions, and Lodging Industry professionals. Two additional certificates will target the cruise industry along with travel intermediaries. The certificates focus on crisis reduction, readiness, response, and recovery efforts from natural- and human-induced disasters that might affect destinations or businesses. Upon completion, participants will receive a Certificate of Participation from the University of Florida.

Not only does UF provide the online institute, but there is also a consultation team to help businesses and institutions assess their needs in regards to tourism and how to proactively prepare for potential impacts from disasters. The aforementioned training opportunities, for creating strategies and procedures, further support the state's vision for emergency management.

Florida's Comprehensive Emergency Management Plan (FCEMP) (2014) includes the Emergency Support Function-18, which focuses on the business, industry, and economic stabilization and whose purpose is to:

> Coordinate local, state and federal agencies and organizations actions that will provide immediate and short-term assistance for the needs of business, industry and economic stabilization. Preparedness and response assistance may include accessing the financial, workforce, technical, and community resources that may affect a community's ability to restore business operations as quickly as possible and resume focus on long-term business strategies. Coordination of local, state and federal business assistance is done primarily through networks of local and regional economic, workforce and tourism development partners, as well as business support organizations who determine the most efficient and effective ways to manage the access to these services at the local and regional level. ESF 18 will support the State Emergency Response Team's (SERT's) efforts via identification and solicitation of resources to meet identified needs. ESF 18 will also support SERT efforts by facilitating and coordinating intermediate and long term economic impact statements.

The tasks of the ESF-18, in relation to tourism, include the following operational objectives of preparedness, response, recovery, and mitigation (SERT 2014). Focusing on preparedness, the main objectives include the provision of educational and training opportunities, the encouragement of agency coordination, information disclosure of disaster planning, developing and maintaining a database of emergency coordinators and private associations, identification of financial resources for long-term and immediate recovery, and maintenance and coordination of data networks for expeditious information delivery (SERT 2014).

Regarding response objectives, ESF-18 assists in tracking recovery activities of primary organizations through all phases of the event, maintaining a list of supportive agencies, coordination and dissemination of public information to communities and businesses, coordination of assessment activities, and provision of support to other emergency support functions (SERT 2014).

The objectives revolving around recovery include coordination of assessments, physical damage estimates, comprehensive and long-term economic impact, contacts for public and private sector agencies, and recovery plans. In addition, recovery incorporates identification and training of bank officers for loan programs and resource centers to aid impacted communities. The last set of objectives relates to the mitigation phase of emergency management and facilitates the identification and access of infrastructure sources along with participation of Hazard Mitigation Assessment Team members (SERT 2014).

In addition to the CEMP, Florida's Division of Emergency Management and the Department of Community Affairs (2010) devised a Post-Disaster Redevelopment Plan to assist communities in developing preparation, mitigation, response and recovery strategies and protocols during pre-disaster periods. The tourism industry is an important component to include in the planning process within the potential stakeholder members, economic redevelopment achievement levels, and environmental compliance.

Tourists tend to be apprehensive about planning vacations to a community that has recently experienced a disaster; however, many Florida communities' economies are dependent on tourism and will need to re-establish this revenue stream as soon as possible. Redevelopment strategies should not overlook aspects of the community that draw tourism, whether that be natural attractions, such as the beaches, or entertainment and cultural establishments. Coastal communities will also need to assess whether tourism-based businesses, such as accommodations and service industry establishments, need assistance in understanding land use strategies to reduce vulnerability and finding ways to assist them in rebuilding in a less vulnerable way (DEM and DCA 2010, p. 66).

To further assist economic development and disaster preparation, Enterprise Florida, Inc. (EFI) (2014), a public-private partnership, generated disaster assistance programs in the hopes of becoming a model for the nation. The goal is to assist businesses with their mitigation, preparedness, response and recovery plans. Within these strategic efforts, EFI provides personalized consultations for each of Florida's 67 counties.

These communication efforts are especially important during disaster declarations and evacuation requests. Beginning with disaster declarations, the Governor has the power to request a disaster declaration for the state and establish a state of emergency. This declaration allows the President to know the nature and scope of an emergency situation and allows for a plethora of assistance programs and grants to become available (FEMA 2014). Moreover, the preliminary damage assessment, which aids in the decision-making process, articulates the impact of a disaster on individuals and businesses. The impact is an important aspect for Florida due to the tourist population. If a declaration is given to the Governor, then the Florida Division of Emergency Management Director can act as the state's Coordinating Officer for all mitigation activities. Should an event occur at the local level, then a mayor, city manager, or board of county commissioners can declare a local state of emergency.

In the case of an evacuation, communication with the tourist population becomes critical due to their transient status within the state. For the Florida, the evacuation process is intermingled with hurricane response due to the risk and vulnerabilities of the natural disaster. More specifically, evacuation routes became a critical component for analysis to reduce negative consequences of mass transitions out of impacted areas. In 2006, the Tampa Bay Regional Planning Council for the Florida Division of Emergency Management conducted a study to analyze issues in hurricane response. Post-hurricane behavioral studies conducted along the Atlantic and Gulf coasts illustrate that many people ordered to evacuate will not and, conversely, people who live in site-built homes far outside the coastal areas will pack up and try to "outrun" the storm ("shadow evacuation"). How we quantify this behavior is key to an accurate transportation analysis. The conducted by the Council used the general hurricane evacuation and response model as well as the surveys conducted following the unprecedented 2004 and 2005 hurricane seasons (Tampa Regional Planning Council 2006, p. 6).

According to the survey, the main behavioral responses impacting evacuation tendencies include:

1. Evacuation Rates – The percentage of population in evacuated and non-evacuated areas that will evacuate during a threat;
2. Evacuation Timing – When the evacuation population would leave their residences in response to a hurricane warning, watch, a given evacuation order or recommendation, and landfall;
3. Vehicle Use – The number of vehicles that evacuating households would use for evacuation;
4. Type of Refuge – The percentage of evacuees that will seek public shelter and other types of refuge such as the homes of friends and relatives, hotel/motels and other locations such as churches, workplaces, and second homes;
5. Evacuation Destinations – The location an evacuee travels to in the event of an evacuation. These destinations can include public shelters, homes of friends/ relatives, hotels/motels, and destinations out of the region;
6. Response by Vacationers – The evacuation response by vacationers, including recreational vehicle (RV) park visitors, encompassing evacuation rate, timing, public shelter use, and vehicle use (Tampa Regional Planning Council 2006, p. III–7).

Focusing more on tourists, the survey was not conducted with the intention to analyze the responses of vacationers. However, the council generated a list of behavioral assumptions from research of other vacation destinations. These include: evacuation rates, type of refuge, destinations, vehicle use, and evacuation timing. In regards to Central Florida, the 2008 version of the evacuation survey highlights the tendency for evacuees to travel to friends or family and they receive the majority of their information from the internet (about 62 %) (Tampa Regional Planning Council

Table 11.1 Hurricane evacuation population by level-base scenarios

DeSoto County

Site-built homes	1,959	3,128	4,885	7,059	7,886
Mobile/Manufactured homes	13,270	13,270	13,270	13,270	13,270
Tourists	0	0	0	0	0
TOTAL	**15,229**	**16,398**	**18,155**	**20,329**	**21,156**

Hardee County

Site-built homes	1,850	1,850	2,775	3,700	4,625
Mobile/Manufactured homes	9,592	9,592	9,592	9,592	9,592
Tourists	0	0	0	0	0
TOTAL	**11,442**	**11,442**	**12,367**	**13,292**	**14,217**

Highlands County

Site-built homes	3,580	3,580	7,160	14,323	17,911
Mobile/Manufactured homes	19,656	19,656	19,656	19,656	19,656
Tourists	0	0	0	0	0
TOTAL	**223,236**	**23,236**	**26,816**	**33,979**	**37,567**

Okeechobee County

Site-built homes	6,863	12,105	21,557	21,557	21,557
Mobile/Manufactured homes	16,761	16,761	16,761	16,761	16,761
Tourists	133	177	202	202	202
TOTAL	**23,757**	**29,043**	**38,520**	**38,520**	**38,520**

Polk County

Site-built homes	22,247	44,495	66,742	88,990	111,237
Mobile/Manufactured homes	68,678	68,678	68,678	68,678	68,678
Tourists	0	0	0	0	0
TOTAL	**90,925**	**113,173**	**135,420**	**157,668**	**179,915**

Tampa Regional Planning Council (2008)

2008). Utilizing predictive analysis, the council also created base scenarios for evacuation for 2010 and determined the evacuation population and levels for various counties in Florida (see Table 11.1). Moreover, scenarios were generated for 2015 as well (see Table 11.2).

Supporting the predictive analysis of evacuation population and levels, the council generated Traffic Evacuation Zones to assist in evaluating the best ways to mitigate potential negative impacts should a hurricane occur (see Fig. 11.1).

Exemplifying the strategic integration of communication and social media is Florida Evacuates (2014), which is a mobile application and website dedicated to providing information for shelters and accommodations. In addition, there are links to the weather service, Division of Emergency Management, and other disaster information and assistance agencies. The application for mobile homes not only provides the preliminary information, but also allows for global positioning systems to map out directions to shelters.

In all, the Florida Division of Emergency Management recognizes the importance of having plans and procedures in place to safeguard the image of Florida as

Table 11.2 Base scenarios for 2015

DeSoto County					
Site-built home	2,549	3,942	6,054	8,711	9,654
Mobile/Manufactured homes	13,792	13,792	13,792	13,792	13,792
Tourists	0	0	0	0	0
TOTAL	**16,341**	**17,734**	**19,846**	**22,503**	**23,446**
Hardee County					
Site-built homes	2,571	2,571	3,856	5,141	6,427
Mobile/Manufactured homes	9,592	9,592	9,592	9,592	9,592
Tourists	0	0	0	0	0
TOTAL	**12,163**	**12,163**	**13,448**	**14,733**	**16,019**
Highlands County					
Site-built homes	**4,184**	4,184	8,368	16,739	20,933
Mobile/Manufactured homes	19,656	19,656	19,656	19,656	19,656
Tourists	0	0	0	0	0
TOTAL	**23,840**	**23,840**	**28,024**	**36,395**	**40,589**
Okeechobee County					
Site-built homes	7,491	13,191	24,588	24,588	24,588
Mobile/Manufactured homes	17,563	17,563	17,563	17,563	17,563
Tourists	133	177	202	202	202
TOTAL	**25,187**	**30,931**	**42,353**	**42,353**	**42,353**
Polk County					
Site-built homes	25,132	50,265	75,397	100,530	125,662
Mobile/Manufactured homes	68,111	68,111	68,111	68,111	68,111
Tourists	0	0	0	0	0
TOTAL	**93,243**	**118,376**	**143,508**	**168,641**	**193,773**

Tampa Regional Planning Council (2008)

a safe place for visitors even in a disaster situation. The importance of this issue was seen in the response to the Deepwater Horizon oil spill in the Gulf of Mexico (Governor's Press Office 2010). In response to the crisis, Governor Crist announced a plan for a marketing campaign focused on tourists with funding from BP. $25 million would be given to Visit Florida in the hopes of assuring potential travelers to the desirability of the state, beginning with the Panhandle region.

Visit Florida, Inc., the state's official marketing avenue for tourism, is responsible for enticing visitors to support economic development. In relation to emergency management, Visit Florida is responsible for the provision and distribution of important information before, during, and after an event. The information is mainly provided through their website portal alongside local weather conditions and social media feeds (e.g. Twitter, Facebook postings). In case of mandatory evacuations, this site provides information on available sources of food and shelter, including lodging accommodations.

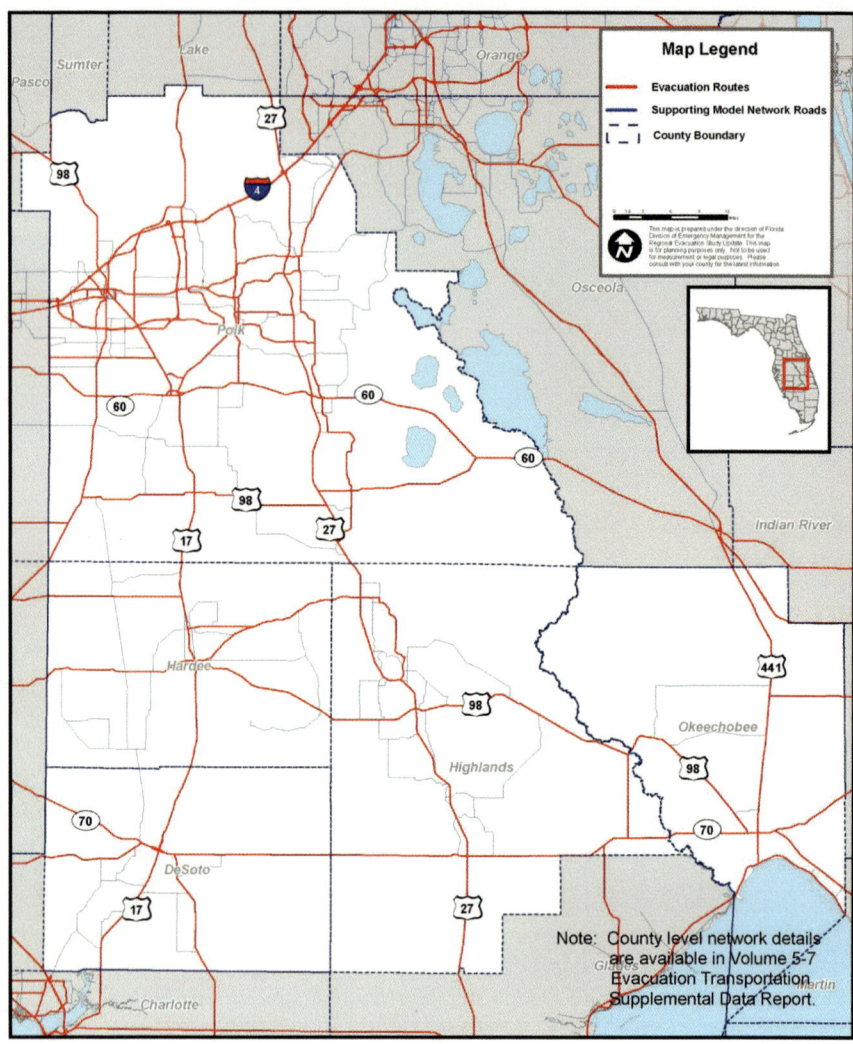

Fig. 11.1 Traffic evacuation routes for Central Florida (Tampa Regional Council Planning 2008)

11.3 Communication and Resilience for Tourism Industry

In Chap. 7, we introduced the adaptive resilience model and argued proactive approaches to increasing resiliency begin with understanding how the area is susceptible to triggers like hazards (i.e. floods, industrial accidents, etc.) and vulnerabilities (i.e. environmental, social, physical and economic conditions)

(Henstra 2010). The discovery of how unique each community is, through examination of aspects like socioeconomics, environmental nuances, and predictability of disaster situations, assists with building capacity. Through holistic planning and strategic management, the impacts of a disaster can be mitigated to reduce and limit severe change resulting from a crisis or disaster (see Fig. 11.2 for example framework) (Ritchie 2004). Adaptive capacity is a way to analyze exposure to risk (e.g. the magnitude and frequency of shocks), sensitivity of the system in responding to given shock or stress, and the ability of involved agencies (including communities, governments, individuals, institutions, organizations, and regions) to anticipate, plan, react and learn from stresses or shocks (Combaz 2014).

As previously alluded to, Florida EM has engaged in the process including tourism in their emergency plans and creating information platforms for visitors to the state during a disaster situation. These aspects support the development of adaptive capacity. However, a recent study exploring disaster planning and preparedness in the tourism industry in Florida found major weaknesses in preparation planning regardless of the fact the majority of travel-related organizations and the top leadership participated in the emergency preparedness. Some of the weaker aspects included low levels of resource allocation, training for employees, access to a centralized reservation system, not having a seat at the emergency operation's center during an emergency, and a lack of communication with FEMA and the national government (Pennington-Gray et al. 2011).

As discussed, there are a variety of strategies in communicating risk in creating resilient communities. The tourism sector requires additional communication strategies since the target population is quite different from the regular citizens (Kapucu et al. 2008). Language, cultural issues, and an understanding of early warning systems can be listed as some challenges for the tourism industry (Collins and Kapucu 2008). The recovery stage deserves careful attention for additional public relations and marketing strategies (Ritchie et al. 2003). In addition, crisis communication strategies, for a quick recovery, requires significant attention along with cross-sector partnership in communities.

11.3.1 Tourism Destination Related Issues

The impacted destination plays an important role in recovering from disasters and building resilient communities. The following factors have been determined to influence the speedy recovery, sustainability, and resilience of an impacted community: publicity and public relations activities; disaster planning and stakeholder engagement; marketing strategies; implementation strategies for comprehensive emergency management plans; level of collaboration among the stakeholders on planning and implementation of planning strategies (including four phases); tourist

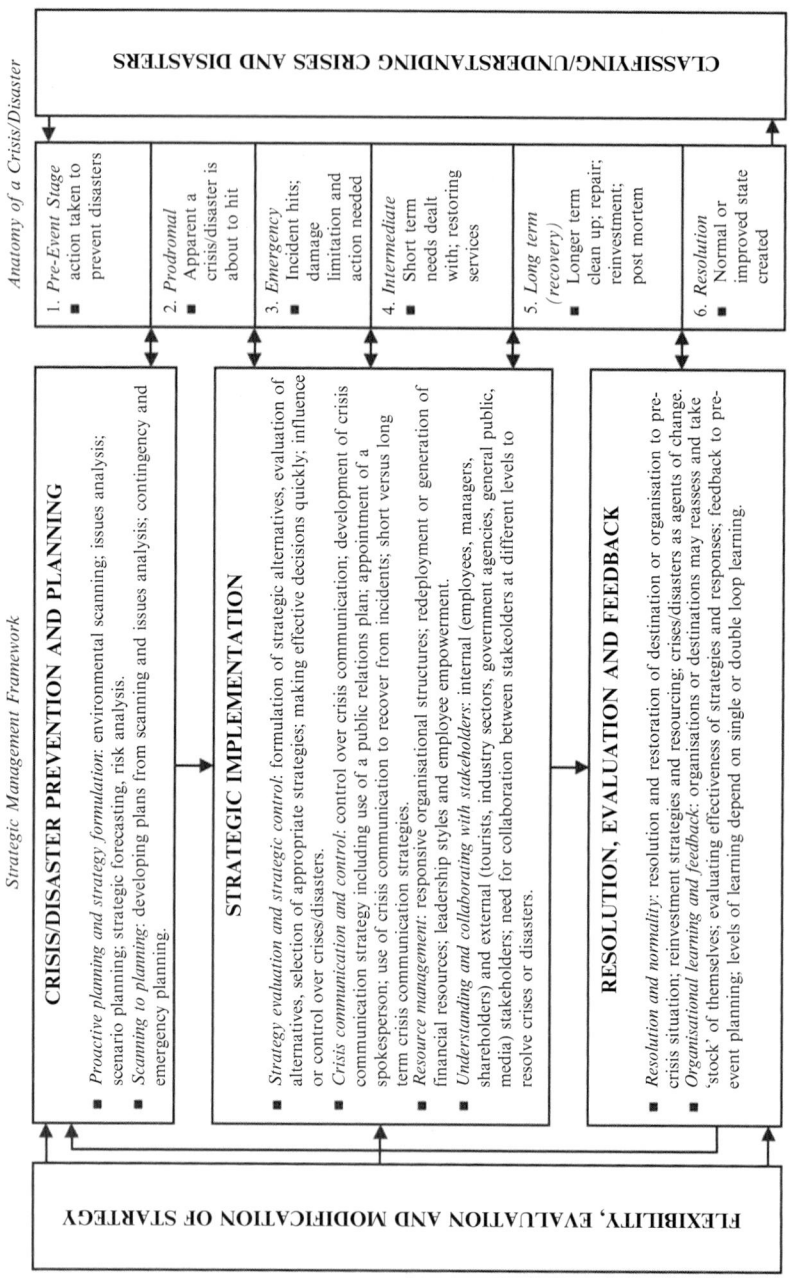

Fig. 11.2 Potential framework for holistic strategic planning (Ritchie 2004)

Anatomy of a Crisis/Disaster

CLASSIFYING/UNDERSTANDING CRISES AND DISASTERS

1. *Pre-Event Stage*
   - action taken to prevent disasters

2. *Prodromal*
   - Apparent a crisis/disaster is about to hit

3. *Emergency*
   - Incident hits; damage limitation and action needed

4. *Intermediate*
   - Short term needs dealt with; restoring services

5. *Long term (recovery)*
   - Longer term clean up; repair; reinvestment; post mortem

6. *Resolution*
   - Normal or improved state created

Strategic Management Framework

CRISIS/DISASTER PREVENTION AND PLANNING

- *Proactive planning and strategy formulation:* environmental scanning; issues analysis; scenario planning; strategic forecasting; risk analysis.
- *Scanning to planning:* developing plans from scanning and issues analysis; contingency and emergency planning.

STRATEGIC IMPLEMENTATION

- *Strategy evaluation and strategic control:* formulation of strategic alternatives, evaluation of alternatives, selection of appropriate strategies; making effective decisions quickly; influence or control over crises/disasters.
- *Crisis communication and control:* control over crisis communication; development of crisis communication strategy including use of a public relations plan; appointment of a spokesperson; use of crisis communication to recover from incidents; short versus long term crisis communication strategies.
- *Resource management:* responsive organisational structures; redeployment or generation of financial resources; leadership styles and employee empowerment.
- *Understanding and collaborating with stakeholders:* internal (employees, managers, shareholders) and external (tourists, industry sectors, government agencies, general public, media) stakeolders; need for collaboration between stakeolders at different levels to resolve crises or disasters.

RESOLUTION, EVALUATION AND FEEDBACK

- *Resolution and normality:* resolution and restoration of destination or organisation to pre-crisis situation; reinvestment strategies and resourcing; crises/disasters as agents of change.
- *Organisational learning and feedback:* organisations or destinations may reassess and take 'stock' of themselves; evaluating effectiveness of strategies and responses; feedback to pre-event planning; levels of learning depend on single or double loop learning.

FLEXIBILITY, EVALUATION AND MODIFICATION OF STARTEGY

education and communication strategies; image enhancement/reputation management programs; and availability of disaster management and crisis response funding (Pizam and Mansfeld 2006; Ritchie 2008).

11.3.2   Image and Perception Management Issues

When disasters impact certain destinations, a negative impact of media coverage (local, national, international) can be expected. Public authorities, in partnering with the stakeholders, can develop strategies in image and perception protection to avoid negative effects of the incidents in economy and resilience of communities. Stakeholders and government authorities can inform the public with accurate and un-biased information. Some of the key elements include, but are not limited to: perceived image of the destination after the disaster; level of risk, hazard, and vulnerability; the effect of media on the image of tourist destination; effect of disasters on travel to destination; effect of the disaster in other similar destinations; and the experience of the destination in similar incidents (Pizam and Mansfeld 2006).

11.3.3   Risk Perception and Communication Issues

The most important element of this issue is pre-disaster planning and stakeholder communication (Murphy and Bayley 1989). Better mitigation strategies and better preparedness lead to effective response and recovery. Prepared/resilient communities are more proactive when responding to disasters (Ritchie 2004). Some other elements for success include: availability of information and communication materials for the tourists in the impacted destinations; availability of contingency plans for each disaster phase; availability of image/perception management and communication plans; multiple information collection and dissemination strategies; availability of public safety resources and personnel; training of employees in the tourism sector; educating local citizens; and collaboration between local community and government leadership (Irvine and Anderson 2006; Kapucu et al. 2008; Murphy and Bayley 1989; Pizam and Mansfeld 2006; Ritchie 2008).

11.4   Focus Group Participant's Perceptions on Tourism

In this section, we analyze the transcripts of our focus group participants from Brevard, Orange, and Osceola counties in relation to tourism. Evaluating the responses from Brevard County, some of the discovered themes include communication, planning, and the tourism industry.

11.4.1 Brevard

11.4.1.1 Communication

Well I know one real big challenge, and that's communication. Uh, we have uh, a call notification system that basically calls telephones, but we can call cell phones. Um, there's no mechanism to make the cell phone companies to allow us to hit their towers and allow us to send out a blanket message and hit all of the cell phones on a tower. And I mean our in-county people that live here, they can go on our website and they can register their cell phones so they can get calls, but all the transient population and all the tourist population, we simply can't talk to them. And so, we're hoping, the technology exists to be able to do that, but the legalities, as it stands right now, you have to have an individual contract with each provider on each tower to be able to do that, and it's just impossible to do and prohibitively expensive. So, to my knowledge, there's no notification system that has that feature. (X: I've actually seen that being tested in NYC). Yeah, but we're hoping to see that, in the next few year you know, to see that.

11.4.1.2 Planning

I think that to your specific question, that is tourism, the first thing that we do is we plan for them. They are parts of our plans, I mean they're on our check lists. As we're going through how we're going to communicate whatever it is that we're gonna communicate, with the methods we have available we make sure that we can reach them. We have a very good relationship with our media, so they're getting a lot our message out. Tourism works very closely with us here, I mean our office of tourism. We do work through social media, we do work with the hotels and motels. We work with the chambers, we work with the cities, especially those on the barrier islands so that they're reinforcing that message. You know, areas that have historically struggled with tourists and tourism after a disaster, the biggest fail point is that they weren't planning for them. We do that well.

11.4.1.3 Tourism Industry

We asked focus groups participant about the tourism industry emergency plans and they shared the following:

Participant A: They have their plans and most of the stuff that they communicate, they'll try to communicate it with the Canaveral port authority. But if it gets really silly, they'll make sure that they communicate directly to the director here, or to the chairman of the county commission if it gets that bad.

Participant B: Cruise lines are probably one of the sectors of the private industry that are pretty well prepared and pretty well tied in.

Participant A: And they're very powerful in their urging.

Participant C: And the nice thing about the cruise industry in general, in addition to them being very aware of their need to prepare, they also have the ability that not a lot of other tourist populations have in that if an area is not safe to go into they can take that entire population and move them somewhere else. So if we're experiencing whatever event, we can communicate with the cruise ships that they have to find alternate ports and they have those plans in place and the ability to do it. We don't have that same luxury with say our motel and hotel guests. You can't just move the holiday inn down the street.

The discussion continued in the following exchange:

Interviewer: And has there even been a situation that they, thinking of cruise ships, have they offered to shelter people?

Participant C: The cruise ship shelter conversation comes up quite a bit and it's not a good idea. It failed during Katrina and it proved that it's not a good idea because it's not designed to be a long term live on board situation. That does not work.

Participant D: In the first place, the great unwashed public tears the ship up before they leave.

Participant C: There is that as well. They just don't have the facilities and if you're talking about putting a lot of people on there for long term, it just, every single facet fails. I mean if you're going to look at that for some sort of short term housing, you're better off looking into your community for vacant rentals, um, vacant lots where you can bring in FEMA trailers. There's a whole lot of other more successful short term disaster housing opportunities than a cruise ship.

11.4.2   Orange County

When evaluating Orange County, the following responses are considered significant in relation to being an in-land county, the provision of information to transient populations, and economic impact.

I don't think we need to. Disney does a great job (laughter). I know people that actually came here during 2004, they made plans to come to Disney, and they got up after Charley came through and there wasn't a single leaf out of place! You know, and I was driving around trees. (Group laughs) I mean, they have great plans.

    Participant A: I mean, I think the thing with tourism is, I mean as for the parks, you know, they're fine. It's the roads though, you know that was their main concern at Sea World, you know, get the roads open so that we can get people in. Because, they do have services that we don't have. I'm not sure who works with the local hotels, but I mean, the tourists that are here are probably better taken care of than our citizens, because those hotels have generators, they have, for the most part, hurricane proof windows, and so they can, the hotels can actually take it; and when our shelters are full, that's where people go.

    Participant B: Actually they go to the hotels before they go to the shelters (laughter) no body wants to go to a shelter and get a peanut butter and jelly sandwich when you can sit around the pool or something, and have a hot shower. We work really well with the hotel industry here, to make sure that we know, our emergency management team, how many rooms are available, so that information is posted and we, we know what the number is. Because then we know how many people are going to have to go to the shelters and how many we need to open to accommodate the people that are comin' in. So, I think we work well with the tourism industry to make sure that that's happening.

11.4.2.1 Challenges

Participant: The other thing that we have to keep in mind is that because we're an inland county, and a very large metropolitan county, we are a destination for not only the tourists, but for evacuees. So we become a host county very quickly, and the problem is because of where we are situated, we can host from the west coast, or east coast, or god forbid, a south Florida evacuation. And we've got to be planning for that, and that's where the tourism industry really comes into play, because that's were most of them are gonna have to go, they're gonna have to go out into the hotels. Um, we need the hotels in the south end of the county, at least, to help us, because we can't, we don't have enough shelter space to shelter them all. So that's the, at the end, why we have such a robust relationship with them.

11.4.2.2 Information/Transient Populations

Participant: I think it's a, it's a typical challenge when you have a large transient population like tourists to provide information, um, because they're partying, they're out there having a good time and they're not sitting in front of the TVs, you know, getting the news. In fact, a lot of them don't watch TV for the week that they're here. So um, there's, uh, there's that challenge. So again you go back to your news organizations, you go back to providing information to the community, to the businesses, so they can disseminate that information to their guests. Um, and uh, have them react accordingly; because they're in a different frame of mind, and I think that's a challenge.

Participant: Yeah, I think they've improved since 04, improved in their willingness to provide information because, before 04, the attitude was it never rains at the theme parks, nothing ever happens in the theme parks, and so they were very concerned about any information that would negatively reflect on that. I think, they've changed their attitude a little bit, they do believe that there is information, disaster information that needs to get out, but they still want to manage it in such a way that it comes across, um, less negatively, or um, (group: laughs – that's the challenge). Their needs are much different than our needs. We want to blitz the community with all of this information, we want them to be as aware as possible, and they want to filter the information so that it's not (other: so that it's a magical world – laughs). So it is a challenge, but we have a good working relationship with them. They are part of our emergency operations plan, uh, we meet with them on a regular basis and the emergency management side of that is very responsive. It's the corporate side, the marketing side, that you have to work with. It's the same with the hotels. The hotel and motel association, you have to really work with them to craft messages so that they're comfortable with how the information is getting out. That's the challenge.

11.4.3 Osceola

In regards to Osceola County, the question of economic impact targeting employment opportunities came to the surface.

11.4.3.1 Employment

Participant: Uh, we have certain neighborhoods that physically are vulnerable. Uh, but I think overall, our general working population is the most vulnerable because we have such a limited, uh, employment focus. And we've tried to change that, uh, the city has tried to change it, the county has tried to change it but we still rely heavily on tourism. And when you do that an one industry impacted by an event then your employees are impacted. People aren't working, you know, it creates a lot of problems.

11.5 Conclusion

In this chapter we discussed some of the pressing challenges and measures put in place by the Florida emergency management structure in relation to disasters and tourism. Due to its geographical and economic reliance on tourism, Florida, like many other tourist destinations most continue to implement and foster conditions promoting disaster resilience. In the face of impingent environmental hazards and disaster situations transient populations, such as tourists, must be accounted and planned for in disaster mitigation plans. Failure to do so can result in catastrophic economic losses, and must importantly can also result in the loss of human lives.

References

Avraham, E. (2006). Public relations and advertising strategies for managing tourist destinations image crises. In Y. Mansfeld & A. Pizam (Eds.), *Tourism, security, and safety* (pp. 233–249). New York: Elsevier.

Combaz, E. (2014). *Disaster resilience: Topic guide*. Birmingham: GSDRC, University of Birmingham.

Collins, M. L., & Kapucu, N. (2008). Early warning systems and disaster preparedness and response in local government. *Disaster Prevention and Management: An International Journal, 17*(5), 587–600.

Drabek, T. E. (1991). Anticipating organizational evacuations: Disaster planning by managers of tourist-oriented private firms. *International Journal of Mass Emergencies and Disasters, 9*(2), 219–245.

Drabek, T. E. (1994a). *Disaster evacuation and the tourist industry*. Boulder: Institute of Behavioral Science, University of Colorado.

Drabek, T. E. (1994b). New study shows that growing tourist industry is inadequately prepared for emergencies. *Hazard Technology, 14*(1), 17–21.

Drabek, T. E. (1995a). Disaster planning and response by tourist business executives. *The Cornell Hotel and Restaurant Administration Quarterly, 36*(3), 86–96.

Drabek, T. E. (1995b). Disaster responses within the tourist industry. *International Journal of Mass Emergencies and Disasters, 13*, 7–23.

Drabek, T. E. (1996). *Disaster evacuation behavior: Tourists and other transients*. Boulder: Institute of Behavioral Science, University of Colorado.

Enterprise Florida (EFI). (2014). *Services: Disaster assistance*. Retrieved from http://www.enter-priseflorida.com/services/disaster-assistance/

Fayos-Sola, E. (1996). Tourism policy: A midsummer night's vision. *Tourism Management, 13*(1), 45–49.

Federal Emergency Management Agency (FEMA). (2014). *The declaration process*. Retrieved from https://www.fema.gov/declaration-process

Florida Division of Emergency Management, & Department of Community Affairs. (2010). *Post-disaster redevelopment plan: A guide for Florida communities*. Retrieved from http://www.floridadisaster.org/recovery/documents/Post%20Disaster%20Redevelopment%20Planning%20Guidebook%20Lo.pdf

Florida Evacuates. (2014). *Mobile app*. Retrieved from http://floridaevacuates.com/index.php

Floyd, M., Gibson, H., Pennington-Gray, L., & Thapa, B. (2004). The effect of risk perceptions on intentions to travel in the aftermath of September 11, 2001. *Journal of Travel & Tourism Marketing, 15*(2), 19–38.

Governor's Press Office. (2010). *Governor Crist announces $25-million tourism advertising campaign: Memorandum of understanding with BP*. Retrieved from http://www.floridadisaster.org/eoc/PressReleases/052510_tourism_mou.pdf

Henstra, D. (2010). Evaluating local government emergency management programs: What framework should public managers adopt? *Public Administration Review, 70*(2), 236–246.

Irvine, W., & Anderson, A. R. (2006). The effect of disaster on peripheral tourism places and the disaffection of prospective visitors. In Y. Mansfeld & A. Pizam (Eds.), *Tourism, security, and safety* (pp. 169–186). New York: Elsevier.

Kapucu, N., Berman, E., & Wang, X. (2008). Emergency information management and public disaster preparedness: Lessons from the 2004 Florida hurricane season. *International Journal of Mass Emergencies and Disasters, 26*(3), 169–197.

Kozak, M., Crotts, J., & Law, R. (2007). The impact of perception of risk on international travelers. *International Journal of Tourism Research, 9*, 233–242.

Law, R. (2006). The perceived impacts of risks on travel decisions. *International Journal of Tourism Research, 8*(4), 289–300.

Murphy, P. E., & Bayley, R. (1989). Tourism and disaster planning. *Geographical Review, 36*–46.

Pennington-Gray, L., Thapa, B., Kaplanidou, K., Cahyanto, I., & McLaughlin, E. (2011). Crisis planning and preparedness in the United States tourism industry. *Cornell Hospitality Quarterly, 52*(3), 312–320.

Pizam, A., & Mansfeld, Y. (2006). Toward a theory of tourism security. In Y. Mansfeld & A. Pizam (Eds.), *Tourism, security, and safety* (pp. 1–27). New York: Elsevier.

Pizam, A., Tarlow, P., & Bloom, J. (1997). Making tourists feel safe: Whose responsibility is it? *Journal of Travel Research, 36*(3), 23–28.

Ritchie, B. W. (2004). Chaos, crises and disasters: A strategic approach to crisis management in the tourism industry. *Tourism Management, 25*(6), 669–683.

Ritchie, B. (2008). Tourism disaster planning and management: From response and recovery to reduction and readiness. *Current Issues in Tourism, 11*(4), 315–348.

Ritchie, B. W., Dorrell, H., Miller, D., & Miller, G. A. (2003). Crisis communication and recovery for the tourism industry: Lessons from the 2001 foot and mouth disease outbreak in the United Kingdom. In C. M. Hall, D. J. Timothy, & D. T. Duval (Eds.), *Safety and security in tourism: Relationships, management, and marketing* (pp. 199–216). Binghamton: The Haworth Hospitality Press.

State Emergency Response Team (SERT). (2014). *Florida's comprehensive emergency management plan.* Retrieved from http://floridadisaster.org/documents/CEMP/2014/2014%20 State%20of%20Florida%20CEMP%20Draft.pdf

Tampa Regional Planning Council. (2006). *Regional behavioral analysis summary.* Retrieved from http://www.sfrpc.com/SRESP%20Web/Vol1-11_ChIII.pdf

Tampa Regional Planning Council. (2008). *Central Florida regional behavioral analysis summary.* Retrieved from http://www.cfrpc.org/programs/statewide-regional-evacuation-study/

Visit Florida Research. (2014). *Research: The power of Florida tourism.* Retrieved from http://visitflorida.org/tools-resources/research/

Chapter 12
Conclusion

Abstract In this chapter we briefly reflect and examine the research questions that were used to guide the research study for the book: How can the concept of resilience be used as a framework to investigate the conditions that lead to stronger, safer, and more sustainable communities? What factors account for the variation across jurisdictions and geographic units in the ability to respond and recover from a disaster? How does the recovery process impact the social, political, and economic institutions of the stricken communities? How do communities, especially rural ones, collaborate with multiple stakeholders (local, regional, state, national) during the transition from recovery to resilience? Can the collaborative nature of disaster recovery help build resilient communities? In addition, we summarize and discuss the policy, practical, and theoretical implications of the book. This chapter discusses how the book findings bring new evidence and insights to the study of disaster resilience, how it integrates knowledge from sociology and public policy and governance to the study of natural disasters, and how the book provides useful and accessible insights not only to academic circles, but also to readers in government, nongovernmental organizations, and the private sector.

Keywords Resilience • Disaster response • Disaster recovery • Rural communities • Stakeholders • Collaboration • Policy • Governance • Management • Florida

12.1 Research Questions

In the preface, the following questions were used to frame the content of the book. Based on the findings, we answer each question below:

How can the concept of resilience be used as a framework to investigate the conditions that lead to stronger, more sustainable communities?

Resilience is a concept that has been incorporated in all emergency management plans at all the local, state, and federal levels (see Chap. 6). Our view of resilience as "the ability to adapt through the redevelopment of the community in ways that reflect the community's values, and goals, and its evolving understanding of external forces with which it must contend" (Kapucu et al. 2013, p. 220) suggests that it

© Springer International Publishing Switzerland 2015   169
F.I. Rivera, N. Kapucu, *Disaster Vulnerability, Hazards and Resilience*,
Environmental Hazards, DOI 10.1007/978-3-319-16453-3_12

can indeed lead to stronger and more sustainable communities. Indeed, the analyses of the perceptions of our study participants on the concept of resilience demonstrated a keen understanding of the conditions that need to be in place for order to promote stronger and more sustainable communities. Examples of these include an understanding of the need for collaborative efforts between the private, public, and non-profit spheres, the need for adequate funding, and constantly developing emergency plans with the flexibility to expect the unexpected. Knowing the social and geographical vulnerabilities of the community can help build resilient communities, not only by understanding the needs of the community, but by also considering that resilience is not only bouncing back, but actually leaping forward. As we suggest in Chap. 6, disaster policies and legislation must undergo a transformative process to become more proactive and predictive in its focus while understanding the importance of context.

One of the most important ways to reduce the effects of a disaster and increase the community resilience is to closely identify and examine all the vulnerabilities to various hazards. This includes looking at the geography, infrastructure, demography of its citizens, as well as historical events. Emergency administrators must expand their vision when evaluating vulnerable groups (aside from the elderly and homeless) and must take into account other issues such as: cultural differences, language barriers, low socioeconomic status, communication issues, and limited resources.

The study of resilience has allowed researchers to come up with different tools to measure a community's social context and vulnerabilities (see Chap. 6) that can be used to implement or reshape governance practices that work for or hinder the sustainability of communities. There are several evaluation methods to identify disaster vulnerability, including the community systems impact tool, social vulnerability indexes, and adapted disaster impact models. Disaster vulnerabilities can be reduced through pre-event activities, such as hazards assessments, land-use regulations, building code development, adoption, and enforcement, warning systems, regular exercises, and education and training programs for communities and public at large.

Context, disturbance, capacity to respond, and reaction are particularly important for disaster resilience.

The Florida Comprehensive Emergency Management Plan emphasizes, similar to FEMA's whole community approach, the need to involve all community members, nongovernmental organizations as well as the private sector to strengthen resilience in accordance with the Presidential Policy Directive (PPD-8). Not surprisingly, when asked to define resilience the open-ended responses from the survey participants closely related to the disaster functions detailed in their CMPs, such as preparedness, "bouncing-back," and recovery.

The concept of resilience has also promoted the developed of different frameworks at the national and international levels that have identified several elements essential to disaster resilience. In particular, the Adaptive Resilience and Community Capital framework (see Chap. 7) which incorporates different elements of community capital and others that are vital in understanding the processes by which resilience (in this case, adaptive resilience) can lead to stronger and more sustainable communities. This framework is a useful way to understand the challenges and barriers to disaster resilience identified in our study including open-ended responses that emphasized lack of funding, complacency, and apathy from the community. These issues were discussed and amplified in our focus group results, alongside other issues including: lack of funding and the economic downturn, complacency/apathy from the community members, brain drain and people leaving, and issues with inadequate facilities and shelter capacity. Other issues identified in Central Florida Counties included: lack of volunteers: issues with transient communities like snowbirds (Brevard); undocumented migrants, personal accountability, and mistrust of government (Orange); the need for coordination between local government agencies (Osceola); disaster resilience issues with the training of volunteers and other emergency management personnel (Seminole); and the distance among people (Sumter).

What factors account for the variation across jurisdictions and geographic units in the ability to respond to and recover from a disaster?

We identified several factors that account for the variation across jurisdictions and geographic units in the ability to respond to and recover from a disaster. These included differential exposure to natural hazards such as hurricanes and tornados, man-made disaster such as nuclear power plant accidents, hazardous materials incidents, mass communication failures, major power disruptions, and oil spills, among others. The ability to respond to and recover from these hazards varies by geographical location whereby resources and impetus is placed on densely populated areas and economic revenue sectors such as themes parks and other touristic destinations. These areas are able to respond quickly and benefit from large emergency management infrastructure and resources. Furthermore, the private sector (particularly the tourism sector) works diligently to return to pre-disaster conditions. The consequence of this recovery vision is that it provides significant recovery challenges for those communities, particularly rural communities, which are sparsely populated and are not viewed as vital economic engines (see Chap. 9 for more details). Take for example, the situation discussed in Chap. 9 where coastal residents drove inland and depilated the gasoline supply of some rural areas, leaving them shorthanded in their recovery efforts.

Another factor for variation in respond and recovery include emergency plans that emphasize hurricane recovery and preparedness (understandable since Florida is prone to them) which favor coastal areas and downplays the recovery efforts from other equally devastating natural hazards such as wildfires, droughts, and tornadoes. Granted, there are guidelines and plans put in place to deal with these types of hazards, but the main emphasis is still place on hurricanes. One quote from the focus groups data summarizes this factor: "Well, and I think a lot of times that we focus, not us as a group, but people, focus only on hurricanes, not all hazards" (see Chap. 4 for details).

Finally, there are variations in social capital, vulnerability, and demographics that vary by jurisdiction. These variations are important in the respond and recovery processes. In some communities, emergency managers must contend with transient populations (including tourists and seasonal residents); others must contend with a large influx of non-English speaking populations, and others with spatially-isolated communities. As we stated before, all disasters are local and the expectations for recovery place burdens on local emergency managers. As highlighted in Chap. 7 it is important to achieve a sense of resiliency, particularly ensuring a high level of support from responsible agencies and political leaders (regardless of jurisdiction) in order to effectively response and recover from disasters.

How does the recovery process impact the social, political, and economic institutions of the stricken communities?

The Florida emergency management system learned from previous disaster response and recovery efforts resulting in enabling and adopting polices plans considered the best in the nation. In Florida there are multiple hazards including hurricanes, tornados, wildfires, lighting, floods, and man-made disasters, several challenges surfaced for hazards in rural communities including geographic, social, economic and political disconnects. These can be overcome with a better understanding of response behavior on a holistic level by incorporating behavioral, social, and political approaches and by mitigation, being prepared, having an effective response, and being able to recover. Building local capacity is accomplished through engaging the citizenry in policymaking and planning, connecting infrastructure performance to resiliency goals, discussing the risks and educating the area, and incorporating sound land-use planning practices, building codes and standards into existing hazard mitigation plans. Overall, hurricanes, tornadoes, floods, and wildfires were identified as recurring hazards that communities in the seven Central Florida counties we studied continue to experience and prepared for. Man-made and technological hazards such as terrorism and damaged bridges are included in comprehensive emergency management plans and are part of routine exercises.

How do communities, especially rural ones, collaborate with multiple stakeholders (local, regional, state, national) during the transition from recovery to resilience?

In a majority of rural communities in Florida, disaster response operations are often handed through volunteers. One of the emerging themes from the focus groups was the perception rural is seen differently in Florida than in other areas in the United States (e.g. rural Midwest). The inland location of the rural counties in our study region was perceived as a particular element providing some insights on a few of the issues particular to rural communities. Another theme identified was the perception isolation of some in rural communities. Although our focus group participants identified several constraints, they also perceived a strong sense of community capital, particularly the perception of the self-sufficient nature of rural residents. The influx of older populations moving into suburban developments within rural areas is particular issue of concern for emergency managers. One of the biggest developments of this kind is The Villages, located in Sumter County.

The analysis of focus group data with farmworkers in Central Florida revealed three major themes: past disaster experiences, self-organizing collective action, and challenges to self-organizing collective action and resilience. The results indicate disaster experiences can serve as a pathway to disaster resilience. We also found that this community of farmworkers was able to utilize their previous disaster recovery experience to create a network of groups to collective take measures to actively prepared and plan for a disaster event. Even though, significant challenges and barriers continue to be present including language issues, anti-immigrant sentiment, relations with the police and farm owners, the reliance on volunteers, VCOEM, and the lack of work at the aftermath of a disaster. These are constant reminders of the vulnerabilities and challenges migrant farmworkers continue to face, not only in Florida, but also throughout the United States.

Can the collaborative nature of disaster recovery help build resilient communities?

Florida, after recovering from the devastation of Hurricane Andrew in 1992, learned that collaboration was essential to help build resilient communities (see Chap. 2 for details). Multi-sector collaborations are imperative to creating a comprehensive, multi-hazard disaster plans based on actual needs. This was taken to heart by former Orange County chairman Richard Crotty as he stated that local mitigation strategies are vital to crease a disaster-resistant (American City and County 2001). As we saw in the aforementioned case in Volusia County, collaboration between different sectors of the community (public, private, and non-profit) can help build resilient communities.

There are several elements that need to be present for collaboration to be effective in building resilient communities including leadership, trust, respect and continual interactions. The collaboration of different sectors in society has being recognized as essential to disaster management and recovery. Approaches like FEMA's "whole community" (Federal Emergency Management Agency 2011) promote collaboration in order to have disaster resilience communities.

12.2 Lessons and Implications

Four major lessons about disaster resilience can be derived from the overview of the key issues of the book chapter. These lessons include: hazards, vulnerability, and resilience concerns; principles of collaborative and integrated approaches for resilience; policy learning; and management and governance practices.

12.2.1 Hazards, Vulnerability, and Resilience Concerns

The first lesson learned deals with the realization that both environmental and social changes have the potential to create new hazards and increase vulnerabilities, particularly for rural communities. As populations continue to move to urban areas, those in rural areas experience considerable financial and social costs, hindered their availability to be disaster resilience. Older populations, undocumented farmworkers, a less diversified economic base, and fewer financial resources if left unattended, have the potential to create an array of challenges and barriers to disaster resilience. Although we found some support for collective action efforts in a group of farmworkers, the perceptions of emergency disaster personnel should serve as warnings of a brewing unfavorable situation for disaster resilience.

Increasing funding and awareness of the issues plaguing rural areas must be at the forefront of any disaster resilience plans, policies, and frameworks. Funding based on population characteristics can limit resources available to rural communities, which can affect the economic base and individual assistance, as well as fragment certain groups of the population. Dispersed communities mean relief arrives to different areas at different times leaving some waiting for additional response and relief provisions. Migrant populations were identified in the focus groups as being vulnerable because of their apprehension to government officials and the anonymous nature of their communities. Therefore, it can be stated that culture and the exclusion of some social networks from support systems can place certain populations at risk in rurally dispersed areas. It is these social systems in place that create demographic and geographic specific social vulnerabilities that must be considered in order to approach resilience from the whole community perspective.

12.2.2   Principles of Collaborative and Integrated Approaches for Resilience

Collaborative and integrated approaches are essential to community disaster resilience. Rural areas, in particular, need to approach disaster management from the viewpoint of creating and sustaining partnerships and building organizational capacity to carry them through mitigation, pre-planning, and response to an effective recovery status. This is critical as the economic base in rural areas is often defined by the ability to sustain agriculture activities which can quickly become disabled after a disaster. Reaching out and extending community partnerships, especially with urban communities, will be a key component to disaster resilience for the whole community. Enhancing the partnerships between urban and rural communities will help increase resources and resilience in the rural setting. Realizing this need prior to a disaster, for example in the pre-planning phase by utilizing group trainings and exercises, can increase the effectiveness of partnerships in the response and recovery phase. The research highlights a need for agencies across geographical and jurisdictional boundaries to work together to reduces hazards and vulnerabilities. Shared resources between rural and urban communities are found to be a key component in reducing vulnerability and increasing recovery time. The issue of not having redundant resources is a problem for rural communities as it can cause dispersion of services and limit the ability to respond in an efficient manner.

12.2.3 Policy Learning

Enacted policies need to be politically and financially supported. It is unreasonable to expect emergency disaster management personnel to adequately perform their duties with shrinking budgets and resources. As we learned in Chap. 6, funding issues were identified as a central barrier to disaster resilience. These compound with mistrust of governmental officials and public complacency are potential policy targets that need to be acknowledge in order to foster disaster resilience. The practice of waiting for a disaster event to occur to implement new policies and procedures is not advisable. Sadly, some of the identified policy issues identified at the aftermath of Hurricane Katrina were painfully repeated during Hurricane Sandy, including evacuation procedures and preparedness/recovery efforts. Central Florida is known by its capacity to implement national and state policies effectively. The region is also known as pioneer to develop policies in building partnerships for building effective disaster response and recovery mechanisms. For example, Central Florida disaster management community developed a local disaster recovery partnership framework before the National Disaster Recovery Framework announced in 2011.

12.2.4 Management and Governance Practices

Management and governance practices can benefit from understanding the perceptions of emergency disaster personnel discussed in this book. The integration of these perspectives is vital to effective disaster management and governance practices as they touch upon the realities of those at the forefront of emergency management. In addition, proper training and interaction with multiple public, private, and nonprofit sector partners (which can be done through the inclusion of multi-agency and multi-jurisdictional representatives) officials will not only mismanage resources but may not visualize their full potential as well. Addressing the problems of funding and time through the use of multi-purposeful trainings will in turn increase the organization's capacity to utilize material and human resources, thus contributing to long term resilience and the effective utilization of partnerships. Having a certain level of reliance on and interaction between partner agencies helps strengthen everyday operations by encouraging the growth of new partnerships, as well as presenting specific community needs to others who may have the resources to help build the capacity of an organization. Central Florida communities give special attention to local capacity, partnerships for disaster response and recovery, and bottom up approaches in managing disasters and crises. Central Florida disaster management community is known by its collaborative decision making in disaster management. We have observed the successful implementation of the partnership perspective during the four hurricanes in 2004 within less than 5 weeks timeframe. The community is also known, if needed, by timely decision-making and coordination with the state and federal government, especially, FEMA.

This book provides an overview of the Florida emergency management system that has been identified as a model for the entire United States (Waugh 2006). It provides a comprehensive review of the current debates surrounding the study of resilience, from federal frameworks, state plans and local initiatives. It also explores and reviews the different evaluation tools to identified vulnerabilities and hazards risks. Most importantly, the book provides first-hand accounts of county emergency managers, nonprofit, and community groups in relation to different issues including vulnerability, hazards, and resilience in rural communities. As stated before, these perceptions are vital to truly understand a community's social and economic context, which are often neglected or not explicitly address in resilience plans and frameworks. The book also provides a lengthy exploration of the issues currently faced by emergency management personnel in rural communities. Finally, the book provides valuable lessons for those interested in building disaster resilience for rural communities and beyond.

References

American City and County. (2001). Disaster preparedness: Cities, county cooperate in hazard plan. *Issues & Trends*, 12–14.

Federal Emergency Management Agency (FEMA). (2011). *Government budgets: Long term trends and drivers and their implications for emergency management*. Washington, DC: FEMA.

Kapucu, N., Hawkins, C., & Rivera, F. (2013). *Disaster resilience: Interdisciplinary perspectives*. New York: Routledge.

Waugh, W. L., Jr. (Ed.). (2006). Shelter from the storm: Repairing the National Emergency management system after Hurricane Katrina. Special issue of *The Annals of the American Academy of Political and Social Science, 604*, 256–272.

Chapter 13
Appendices A, B and C

Abstract This chapter includes the survey instrument and focus group interview script utilized to collect some of the data discussed in the book. In addition, there is a content analysis of news articles from the 2014 hurricane season with regards to hurricane preparedness, resilience, vulnerability, and hazards.

Keywords Survey • Focus groups • Interview script • News articles • Hurricane preparedness • Hurricane resilience • Hurricane vulnerability • Hurricane hazard • Florida

13.1 Appendix A

13.1.1 *Building Disaster Resilience and Sustainability in Central Florida*

This survey helps to delineate factors that are important to create disaster resilient communities. This survey will be used to identify Central Florida counties' current level of resilience to disasters and how it can be improved. The survey takes about 20–30 min to complete. Your responses are confidential, and will not be revealed without your consent; only aggregate results will be made available. We would be happy to provide you with the final results upon request.

Thank you very much for your cooperation

© Springer International Publishing Switzerland 2015   177
F.I. Rivera, N. Kapucu, *Disaster Vulnerability, Hazards and Resilience*,
Environmental Hazards, DOI 10.1007/978-3-319-16453-3_13

Contact: Professor Naim Kapucu
Department of Public Administration
University of Central Florida
HPA II, 238
Orlando, FL 32816-1395
(407)823-6096
FAX (407)823-5651
kapucu@ucf.edu

Please tell us about yourself:

Are you the addressee?

[ ]   Yes
[ ]   No → Please state your position/title here: _____
[ ]   Organization representing: _____

How familiar are you with emergency management & planning in the counties that you (or your organization) primarily serve? (Please check one)

[ ]   Very Familiar
[ ]   Familiar
[ ]   Somewhat familiar
[ ]   Unfamiliar

Which counties does your organization serve? (Check all that apply)

Part 1: This section focuses on disaster preparedness for your organization (PREPAREDNESS)

[ ] Brevard   [ ] Citrus   [ ] Flagler   [ ] Lake   [ ] Levy   [ ] Marion
[ ] Orange   [ ] Osceola   [ ] Seminole   [ ] Sumter   [ ] Volusia

Question 1: Please assess the following statements regarding the level of preparedness of your organization. Please use the following scale:

Strongly Agree	Agree	Neither Agree Nor Disagree	Disagree	Strongly Disagree
5	4	3	2	1

[ ]   We conduct pre-season coordination meetings with local community organizations.
[ ]   We make emergency communication procedures available to our stakeholders.
[ ]   We regularly conduct (e.g. quarterly) emergency training and exercises in our community to create awareness.
[ ]   My organization has data backup to use for maintaining functions in case of a disaster.
[ ]   My organization rehearses our disaster response plan regularly.
[ ]   We market relevant parts of our plans (such as lockdown, shelter in place, evacuation, etc.) through city/town hall meetings.

[ ]   Our staff avail FEMA Independent Study certification courses.
[ ]   Key staff is trained in ICS (Incident Command System) and NIMS
      (National Incident Management System) courses.
[ ]   My institution has adequate resources to implement and maintain comprehensive
      training programs.
[ ]   Pre-drill and post-drill surveys are used to update plans and procedures.

Question 2:   How often are the following exercises or drills conducted by your institution? Please use the following scale:

Quarterly	Bi-annually	Annually	two years	Never
5	4	3	2	1

[ ]   Tabletop exercises
[ ]   Functional drills/stimulations
[ ]   Full-scale exercises
[ ]   Evacuation site visits

[ ]   Emergency Operation Center (EOC) visits

Part 2: This section focuses on the role of inter-organizational networks on building disaster resiliency. (INTERORGANIZATIONAL NETWORKS)

Question 1:   Please assess the following statements regarding the development of interorganizational networks for (or by) your organization. Please use the following scale:

Strongly Agree	Neither Agree Agree	Nor Disagree	Strongly Disagree	Disagree
5	4	3	2	1

[ ]   My organization is involved in partnerships for disaster management with other organizations.
[ ]   My organization is involved in making decisions for disaster management.
[ ]   My organization shares resources with other organizations for E.M. purposes because
      we need their resources for reaching organizational goals.
[ ]   My organization in the network share resources with others for reaching overall network goal
[ ]   My organization's management considers stakeholders' interests in developing
      organizational strategies.
[ ]   My organization's stakeholders have a role/influence on our organization's strategies.
[ ]   My organization considers itself as a steward for its stakeholders (e.g. community, clients,
      other organizations).
[ ]   Organizations that we have relations with periodically contact each other to discuss issues
      pertaining to emergency management.
[ ]   Organizations that we have relations with have developed long-term relationships
      among each other.
[ ]   My organization signs interorganizational agreements (e.g. MOUs) with our partners
      to enhance our long-term collaboration.
[ ]   The more organizations in our network sustain their relationships across time, the more
      effective they manage disasters.

Question 2:   Please assess the following statements regarding the sustainability of your organization's network. Please use the following scale:

Almost Always	Almost Frequently	Sometimes	Seldom	Never
5	4	3	2	1

[ ]   Organizations in our emergency management network regularly discuss E.M. issues.
[ ]   Organizations in our emergency management related network develop long-term relationships among each other.
[ ]   In the absence of disasters, organizations are involved in collaborative practices (such as exercises, drills).
[ ]   The organizations in our network collaborate in the absence of disasters.

Please assess the following statements according to the scale below

Strongly Agree	Neither Agree Agree	Nor Disagree	Strongly Disagree	Disagree
5	4	3	2	1

[ ]   Our emergency management related network sustains inter-organizational relationships to improve disaster preparedness and management in the future.
[ ]   The more organizations in our network sustain their relationships across time, the more effective they manage disasters.
[ ]   The success of our emergency management related network is dependent on the strength of inter-organizational relationships.

Question 3:   Please assess the following statements that are important for information communication technology utilization, information sharing, and communication for your organization with respect to disaster management. Please use the following scale:

Strongly Agree	Neither Agree Agree	Nor Disagree	Strongly Disagree	Disagree
5	4	3	2	1

[ ]   We focus on emergency management information sharing with outside organizations.
[ ]   My organization exchanges information with other organizations.
[ ]   Organizations in our network constantly communicate and exchange information.
[ ]   The interorganizational network's operations are streamlined by technological tools of communication and coordination.
[ ]   Organizations in the network have a sufficient technical capacity for emergency management.
[ ]   The use of information and communication technology (e.g. internet tools) facilitates the operations of the network.
[ ]   Inter-organizational operations in the network are supported by emergency/disaster information management systems.
[ ]   The network would be less efficient without a technological capacity used for communication and coordination.
[ ]   If our emergency management network is effective, it is mainly due to the use of information and communication technologies.

Question 4: **Please assess the following statements regarding learning in inter-organizational networks that your organization is involved in. Please use the following scale:**

Strongly Agree	Neither Agree Agree	Nor Disagree	Strongly Disagree	Disagree
5	4	3	2	1

[ ]   My organization learns new knowledge about emergency management from its relations with partners.

[ ]   My relationship with partnering organizations helps us to improve our knowledge to conduct emergency management responsibilities.

[ ]   My organization constantly shares emergency management experience and best practices with its partners.

Part 3: This section focuses on organizational capacities for building disaster resiliency (organizational knowledge and learning) (ORGANIZATIONAL CAPACITY)

Question 1: **Please assess the following statements for your organization regarding organizational knowledge and expertise. Please use the following scale:**

To a Great Extent	Don't know/ Somewhat	Can't say	Not at Very little	All
5	4	3	2	1

[ ]   Top level managers in our organization are supporting E.M. efforts in our county.

[ ]   We successfully implemented our E.M. plan during a disaster.

[ ]   We learned about our problems by assessing previous disasters.

[ ]   Compared to pre-disaster conditions, my organizational disaster preparedness knowledge has increased after each disaster.

[ ]   Compared to pre-disaster conditions, my organizational disaster response knowledge has increased after each disaster.

[ ]   My organization has used lessons learnt from previous disasters to prepare for and mitigate against future disasters.

[ ]   My organization has used lessons learnt from previous disasters for better response to disasters.

Question 2: **Please select the most appropriate answer for the following statements about your organization's emergency management expertise and experience.**

Our organization has been dealing with emergencies for (years)

[ ] 0–2   [ ] 3–5   [ ] 6–8   [ ] 9–11   [ ] more than 11

The number of disasters we have experienced is

[ ] 0–2   [ ] 3–5   [ ] 6–8   [ ] 9–11   [ ] more than 11

The number of E.M. certifications we have received is

[ ] 0–2   [ ] 3–5   [ ] 6–8   [ ] 9–11   [ ] more than 11

We have had a continuity of operations plan for (years)

[ ] 0–2   [ ] 3–5   [ ] 6–8   [ ] 9–11   [ ] more than 11

Part 4: This section focuses on the physical environment and planning aspects of disaster resiliency (planning for sustainable land use and environment, land use characteristics and mitigation, planning process and plan quality) (PHYSICAL ENVIRONMENT AND PLANNING)

Question 1: **To what extent has your organization been involved in the following to improve disaster resiliency in the communities you primarily serve? Please use the following scale:**

Almost Always	Frequently	Almost Sometimes	Seldom	Never
5	4	3	2	1

[ ]   Actions oriented towards our ecological systems and landscape
[ ]   Actions oriented towards social support systems
[ ]   Actions oriented towards economic development and diversity
[ ]   Actions oriented towards equitable access to resources for all citizens within the community
[ ]   Actions oriented towards housing
[ ]   Actions oriented towards transportation
[ ]   Actions oriented towards energy
[ ]   Actions oriented towards land use and design
[ ]   Actions oriented towards public facilities
[ ]   Other (Please specify):

Question 2: **Which of the following *disaster mitigation strategies* has your organization been involved in developing for the communities that you primarily serve? Please use the following scale:**

Almost Always	Almost Frequently	Sometimes	Seldom	Never
5	4	3	2	1

[ ]   Planning Tools (e.g. land acquisition, floodplain management, environmental review)
[ ]   Zoning Tools (e.g. performance standards, special use permits, density controls)
[ ]   Subdivision Controls (e.g. water supply, road access/width)
[ ]   Design Controls (vegetation, design review, building codes)
[ ]   Financial Tools (e.g. relocation aid; special districts; lending policies)
[ ]   Management Tools (e.g. inter-jurisdictional coordination; public education)
[ ]   Other (Please specify):

Question 3:   To what extent has your organization participated in the following for the counties that you primarily serve? Please use the following scale:

Almost Always	Frequently	Sometimes	Almost Seldom	Never
5	4	3	2	1

[ ]   Educational forum to discuss land use planning
[ ]   Community land use visioning workshop
[ ]   Consensus building workshop
[ ]   Community needs assessment survey
[ ]   Conflict resolution
[ ]   Citizen boards and commissions
[ ]   Community risk assessment
[ ]   Web-based information exchange
[ ]   Other (Please specify):

Question 4:   Based on the items listed below, how would you assess the comprehensive planning *process* for the counties that your organization serves? Please use the following scale:

Excellent	Good	Neither good nor poor	Very Poor	Poor
5	4	3	2	1

[ ]   Community-wide
[ ]   Educated community members on hazards and mitigation
[ ]   Participatory and incorporated innovative ways to manage risk
[ ]   Actively sought resident input into planning for emergency management
[ ]   Actively sought business input into planning for emergency management
[ ]   Created a greater sense of place among residents of the community
[ ]   Provided strategies on how to achieve a more economically integrated and diverse population
[ ]   Enhanced the local commitment to hazard mitigation
[ ]   Improved the local capacity of stakeholders to plan for potential natural hazards
[ ]   Improved the feasibility of integrating natural hazard mitigation into the land development market
[ ]   Supported a regional perspective on hazards
[ ]   Developed clear implementation steps
[ ]   Negotiated among competing interests
[ ]   Incorporated "stories" and personal experiences

Question 5: **Please assess the** *quality* **of the comprehensive plan of the counties your organization primarily serves according to items below.** **If you serve multiple counties please answer according to your primary county you serve.** **Please use the following scale:**

	Neither good		Very	
Excellent	Good	nor poor	Poor	Poor
5	4	3	2	1

[ ]   Community level actions are developed locally
[ ]   Plan represents the values of the community
[ ]   Plan provides a range of policy choices at the local level
[ ]   Goals represent community resilience objectives
[ ]   Plan presents facts about hazards and increasing public awareness of them
[ ]   Systematically examines the adequacy of existing hazard mitigation measures being used
[ ]   Represents a clear vision of hazard resilience
[ ]   Presents specific policy goals and objectives for achieving resiliency
[ ]   Reflects a consensus on the need to take action to reduce vulnerability and to find courses
       of action that are politically acceptable
[ ]   Outlines how the community will investigate and use a variety of approaches
       to hazard mitigation
[ ]   Provides guidance to the day-to-day decisions of local officials in approving or disapproving
       development proposals
[ ]   Helps to coordinate the actions of various local government departments
       and non-government organizations that affect vulnerability

Part 5: This section focuses on mitigation for building disaster resiliency (MITIGATION)

Question 1: **Please assess the following statements that are important for mitigation with respect to your organization. Please use the following scale:**

Strongly	Neither Agree		Strongly	
Agree	Agree	Nor Disagree	Disagree	Disagree
5	4	3	2	1

[ ]   My organization is aware of the hazards that create a high level of risk for our community.
[ ]   My organization is aware of its own vulnerabilities to disasters
[ ]   My organization makes plans to reduce its vulnerabilities.

Please answer the following items according to scale below

Yes	No
1	2

[ ]   My organization has a role in citizen corps programs.
[ ]   All the communities that our organization serves are part of the storm ready
       communities initiative.

[ ]   My organization has disaster insurance.
[ ]   We have completed all the local mitigation strategy goals that are addressed
      by the county emergency management organization.
[ ]   My organization is located outside of disaster prone areas.

Part 6: This section focuses on indicators of disaster resiliency

Question 1: Please assess the following statements that are indicators of response effectiveness with regard to your organization. Please use the following scale: (*RESPONSE EFFECTIVENESS*)

Strongly Agree 5	Neither Agree Agree 4	 Nor Disagree 3	Strongly Disagree 2	 Disagree 1

[ ]   My organization is aware of its partners for cooperation in response to disasters.
[ ]   In case of a disaster, our organization is able carry out its disaster response according
      to a disaster response plan.
[ ]   My organization is flexible enough to handle unexpected conditions while responding
      to an incident.
[ ]   My organization's responsibilities for response to an incident are clearly defined.
[ ]   My organization's responsibilities for response to an incident are clearly structured.
[ ]   My communication tools are operable with other organizations' communication tools.

Question 2: Please assess the following elements that are indicators of recovery with respect to the communities that your organization primarily serves. Please use the following scale: (*RECOVERY*)

Communities can return to routine life in (weeks) after a disaster?

[ ] Less than 2   [ ] 3–4   [ ] 5–6   [ ] 7–8   [ ] More than 8

Communities can return to its routine functions with a cost (% of actual property
   value)

[ ] 0–20 %   [ ] 21–40 %   [ ] 41–60 %   [ ] 61–80 %   [ ] 81 %-100 %

Communities can continue education after a disaster in (weeks)

[ ] Less than 2   [ ] 3–4   [ ] 5–6   [ ] 7–8   [ ] More than 8

Please use the scale below for the following three statements

Strongly Agree	Neither Agree Agree	Nor Disagree	Strongly Disagree	Disagree
5	4	3	2	1

[ ] My organization is aware of the disaster recovery plan goals of the communities that we serve.
[ ] A disaster recovery plan can restore our community to a better condition than pre-disaster time.
[ ] My organization has a role in recovery efforts.

Question 3: Please assess the following elements that are indicators of the adaptive capacity of your organization. Please use the following scale: (*ADAPTIVE CAPACITY*)

Strongly Agree	Neither Agree Agree	Nor Disagree	Strongly Disagree	Disagree
5	4	3	2	1

[ ] If our organization loses its physical resources in a disaster, we have adequate substitute resources.
[ ] My organization has an alternative location to operate in times of disasters.
[ ] My organization can rapidly mobilize its resources for managing an emergency.
[ ] My organizational resources are robust enough to withstand disasters.
[ ] My organization has the capacity to utilize materials and human resources to manage emergencies.

Part 7: This section focuses on social factors impacting disaster resiliency (*SOCIAL FACTORS*)

Question 1: Please assess the following statements that are important for community competence in your community. If you serve multiple counties please answer according to the primary county you serve. Please use the following scale:

Strongly Agree	Neither Agree Agree	nor Disagree	Strongly Disagree	Disagree
5	4	3	2	1

[ ] The community my organization serves is flexible and has adequate problem solving skills.
[ ] If we asked everyone to conserve water or electricity because of some emergency, people in my community would cooperate.
[ ] The community my organization serves is prepared for a disaster situation.
[ ] In the community my organization serves exists a social environment of mutual assistance, caring, and support.
[ ] People in the community my organization serves trust each other.
[ ] People in the community my organization serves trust the government.
[ ] People in the community my organization serves have a strong sense of belonging.

Question 2:   Please assess the following statements with respect to civic engagement in your community. Please use the following scale:

Strongly Agree	Neither Agree Agree	Nor Disagree	Strongly Disagree	Disagree
5	4	3	2	1

[ ]   People in the community my organization serves engage in volunteer work.
[ ]   People in the community my organization serves are active members of a group
      or association.
[ ]   People in the community my organization serves engage in fund raising activities.
[ ]   People in the community my organization serves vote in political elections.
[ ]   People in the community my organization serves contact their elected officials.
[ ]   People in the community my organization serves engage in charitable giving/donations.

13.1.2   Demographic/Socioeconomic Factors

Question 3:   Please assess the following statements about racial and ethnic diversity that are important in your community and organization. Please use the following scale:

Strongly Agree	Neither Agree Agree	Nor Disagree	Strongly Disagree	Disagree
5	4	3	2	1

[ ]   People in my organization are racially and ethnically diverse.

Question 4:   Please assess the following statement of social class in your community. Please use the following scale:

Upper Class	Middle Class	Working Class	Poor	Very Poor
5	4	3	2	1

[ ]   Would you say that the community your organization serves is mostly:

Question 5:   Please assess the following statements of health/mental health status in your community. Please use the following scale:

Excellent	Very Good	Good	Poor	Very Poor
5	4	3	2	1

[ ]   Overall, how would you rate the physical health of the community your organization serves.
[ ]   Overall, how would you rate the mental health of the community your organization serves.

Question 6: Please assess the following statements on gender regarding your organization.

What percentages of people in your organization are: Male [ ] Female [ ] (Please round to 100 %)

Question 7: Please assess the following statement for the special needs populations that are important in your community. Please use the following scale:

Strongly Agree	Neither Agree Agree	Nor Disagree	Strongly Disagree	Disagree
5	4	3	2	1

[ ]   My organization is prepared to serve special needs populations (e.g. tourists, homeless, nursing homes, prisoners, etc.) during a disaster situation.

What percentage of special needs people are in the community your organization serves? [ ]

Question 8: Please assess the following statements of age that are important in your community and organization. Please use the following scale:

0–18	19–49	50–64	Over 65	Don't know
1	2	3	4	5

[ ]   In your estimation, what is the average age of the people working in your organization?

13.1.3   Social Context (Economy and the Media)

Question 9: Please assess the following statements of the economy that are important in your community and organization. Please use the following scale:

Strongly Agree	Neither Agree Agree	Nor Disagree	Strongly Disagree	Disagree
5	4	3	2	1

[ ]   The downturn in the economy has negatively impacted the community your organization serves.
[ ]   The downturn in the economy has negatively impacted your organization.

Please assess the impact of the economic downturn in the community your organization serves in the following areas: (Use the following scale)

Large Impact	Moderate Impact	Don't know/Very Little Can't say	Impact	No Impact
5	4	3	2	1

[ ]   Employment
[ ]   Home ownership
[ ]   Crime
[ ]   Sense of community
[ ]   Trust in government
[ ]   Other (please specify)

Please assess the impact of the economic downturn in the community for your organization in the following areas: (Use the following scale)

Large Impact	Moderate Impact	Very Little Can't say	No Impact	Impact
5	4	3	2	1

[ ]   Staff
[ ]   Equipment
[ ]   Technology
[ ]   Services provided to the community
[ ]   Other (please specify)

Question 10: Please estimate the impact of the news media to your organization. Please use the following scale:

Very Positive	Positive	No effect	Negative	Very Negative
5	4	3	2	1

[ ]   The news media has the following effect on my organization

Part 8: This section includes open ended questions regarding your thoughts on disaster resiliency

Question 1: How do you define disaster resilience?

Question 2: Are there additional elements (that have not been covered in this survey) that you think are important to create disaster resilient communities?

Question 3: What are the obstacles to build disaster resilient communities?

Part 9: This section focuses on organizational demographics (CONTROL VARIABLES)

Which sector are you in?

[ ] Public   [ ] Private   [ ] Nonprofit

How long have you been involved in E.M. field?
How much is your organization's average annual budget for emergency management purposes?
How many staff is currently employed by your organization?
How many different services do you provide to your clients?
What is the primary service your organization delivers?
What percentage of your service area is rural (Rural communities are defined as communities with population less than 10,000)?

Thank you very much for your participation!

13.2   Appendix B

Building Disaster Resilience and Sustainability in Rural Communities in Central Florida
   Focus Groups Interview Script

(A) *Opening*

   1. Introduction of interviewer and purpose of interview. Conversational by-play.
   2. Explanation of informed consent procedures and consent form. Give copy of consent.

(B) *Interview Content*

   1. *Resiliency*

      What is your (or your organization's) definition of disaster resilience?
      How does your organization's mission, resources, and goals impact disaster resilience, particularly in rural communities?
      Discuss some of the challenges your organization is currently facing. How do these challenges affect disaster resilience?
      What is the biggest challenge facing your community right now? How do you think this issue will affect disaster resilience?

Probe for:

Economy
Miscommunication or lack of communication
Diverse populations and needs
Others

What is the strongest asset your community has? Do you think that this benefits disaster resilience? How so?

2. *Mitigation/Preparedness/Response/Recovery*

Who in your community is most vulnerable to a disaster? What can be done to make them less vulnerable?

Who in your community is the most prepared for a disaster? What can be learned from them that could help other community members?

What role should government have in disaster response?

What can nonprofit organizations / local faith-based community organizations do to help members of your community prepare for/recover from a disaster?

What can the local business community do to help members of your community prepare for/recover from a disaster?

3. *Community Vulnerability & Disaster Resilience*

Thinking about "natural disasters," (Give examples of natural disasters: flooding, wildfires, tornadoes, hurricanes, etc.) what do you think is the greatest threat to your community? Do you think the community is prepared for this type of natural disaster threat? What can be done to help?

What threats do disasters pose to the business community? What can be done to help business prepare for disasters?

What threats do disasters pose to the faith-based community? What can be done to help the faith-based community prepare for disasters?

What threats do disasters pose to the non-profit organizations? What can be done to help non-profit organizations prepare for disasters?

What threats do disasters pose to neighborhoods? What can be done to help neighborhoods prepare for disasters?

4. *Community Relations and Adaptation*

Can you tell us a little about how communities (residents, government and non-government organizations) have survived, reconstituted, and adapted in the response and rebuilding stages of disaster recovery?

Please tell us about personal ties among individuals in the context of disasters. In what ways have interpersonal relations improved the capacity to withstand a disaster?

What features of the community have you seen adapt in response to the threats from disasters?

Can you characterize for us how tangible assets of communities and organizations (financial, labor) are shared across communities and organizations to improve resiliency? What can be improved to strengthen these relations?

What about assets that is less tangible, such as trust and shared goals, in the ability of communities to withstand disasters? How do these features of relations improve the ability of communities to adapt in response to the potential threats from disasters?

5. *Social Media/News Media*

What should be the role of local news media before, during, and after a disaster? In what way do media reports on disasters impact your preparedness?

Do you use social media? What role do you think that social media can have before, during, and after a disaster?

Does your organization use social media? What has been your social media strategy? Have you noticed any benefits or difficulties of using social media?

How would you describe your organization's relationship with local news media?

Probe for: Do they feel they can "easily" get their message/image across? Would they like to communicate better but do not have the resources to do so?

6. *Politics and Government Action*

Are there any political stresses in your community that could affect disaster resilience? How do you think this can be mitigated?

What policies should be pursued by local government to improve resiliency? In your assessment, what makes these policies difficult to implement?

7. Special Need Populations (Tourism) (Asked if appropriate for county)

What role does tourism play in your community? How would maintaining tourism be a part of your disaster resiliency plans?

How do you work/plan for with visitors and/or tourists, as compared to community members?

What threats do disasters pose for tourism? What can be done to help tourist industries prepare for disasters?

Do use any media outlets specifically for visitors or tourists?

Probe for how often? Which media outlets? And probe for further discussions of their experiences.

(C) Conclusion

Are there any additional comments you would like to make?

Thanks for your participation.

13.3 Appendix C

13.3.1 *2014 Florida Hurricane Season News Articles*

Hurricane Preparedness

- Senate Environment and Public Works Subcommittee on Clean Air and Nuclear Safety Hearing

- Basic Point: Recapped national catastrophes and the importance and influence of risk insurance
- Hedde, C. (2014, July 30). Senate environment and public works subcommittee on clean air and nuclear safety hearing. *Federal Information & News Dispatch, Inc.* Retrieved from http://insurancenewsnet.com/oarticle/2014/07/30/senate-environment-and-public-works-subcommittee-on-clean-air-and-nuclear-safety-a-537459.html#.U9uGibAg-M8

- Hurricane Season Preparedness Important for Protecting Florida's Environment

  - Basic Point: Preparation is key and essential to preventing environmental harm
  - Florida Department of Environmental Protection (2014, June 2). Hurricane season preparedness important for protecting Florida's environment. *DEP Press Office*. Retrieved from http://content.govdelivery.com/accounts/FLDEP/bulletins/ba9a18

- DCF Prepares for 2014 Hurricane Season with Mock Disaster Exercise More than 100 Employees to Participate in Drill to Ready Food for Florida Program

  - Basic Point: Preparation activity undertaken by the Department of Children and Welfare to better anticipate needs during the 2014 season
  - (2014). DCF prepares for 2014 hurricane season with mock disaster exercise more than 100 employees to participate in drill to ready food for Florida program. *States News Service*. Retrieved from http://www.myflfamilies.com/press-release/dcf-prepares-2014-hurricane-season-mock-disaster-exercise

- Florida Homeowners: Get the Most From Insurance this Hurricane Season Plus 3 Ways to Get Your Home Ready Now

  - Basic Point: Although Florida insurance companies are prepared to handle the impact of hurricanes, residents should not become complacent
  - PRWeb. (2014, June 11). Florida homeowners: Get the most from insurance this hurricane season plus 3 ways to get your home ready now. *Insurancenewsnet.com*. Retrieved from http://insurancenewsnet.com/oarticle/2014/06/11/florida-homeowners-get-the-most-from-insurance-this-hurricane-season-plus-3-way-a-516348.html#.U9uQO7Ag-M8

- Gov. Scott Reminds Floridians to Get a Plan

  - Basic Point: Reminder to Floridians to plan and prepare for the hurricane season
  - Governor's Press Office (2014, July 15). Gov. Scott reminds Floridians to get a plan. WCTV.TV. Retrieved from http://www.wctv.tv/home/headlines/Gov-Scott-Reminds-Floridians-to-Get-a-Plan--267257491.html

- Experts still call for slower season

  - Basic Point: Prediction of one more major storm and then lesser impact storms

- Kaye, K. (2014, July 31). Experts still call for slower season. *Sun Sentinel*. Retrieved from http://articles.sun-sentinel.com/2014-07-31/news/sfl-klotzbachgray-seasonal-update-20140731_1_busiest-stretch-tropical-activity-least-one-major-hurricane

- Hurricane warning: Don't let your home insurer blow you over.

  - Basic Point: Be cautious in the insurance process and be prepared for large claims should a major storm impact your home
  - Blystkal, J. (2014, July 15). Hurricane warning: Don't let your home insurer blow you over. *FoxBusiness.com*. Retrieved from http://www.foxbusiness.com/personal-finance/2014/07/15/hurricane-warning-dont-let-your-home-insurer-blow-over/

Hurricane Resilience

- Public-Private Action on Resilience is Needed- Now

  - Basic Point: Push for action more than theory; time for planning is now
  - Mullen, J. (2014, July 22). Public-private action on resilience is needed- now. *Emergency Management: Disaster Preparedness and Recovery.* Retrieved from http://www.emergencymgmt.com/disaster/Public-Private-Action-on-Resilience.html

- Congress Mulls Various Measures Targeting Disaster Mitigation and Resilience

  - Basic Point: Congress debating more legislation to strengthen building standards and disaster savings efforts for future catastrophes
  - Hofmann, M.A. (2014, July 20). Congress mulls various measures targeting disaster mitigation and resilience. *Business Insurance*. Retrieved from http://www.businessinsurance.com/article/20140720/NEWS04/307209976?tags=%7C306%7C64%7C302

- Wall of Wind Research Aiming to Create a Hurricane Resilient Coastline

  - Basic Point: A wall of wind was created to predict needed building codes
  - Winkley, S. (2014, July 31). Wall of wind research aiming to create a hurricane resilient coastline. *KBTX.com*. Retrieved from http://www.kbtx.com/home/headlines/Wall-of-Wind-Research-Aiming-to-Create-a-Hurricane-Resilient-Coastline---269147061.html

Hurricane Vulnerability

- Poorest Communities are Least Resilient to Weather Extremes- Study

  - Basic Point: Poor communities are affected the most by disasters
  - Hulac, B. (2014, July 28). Economics: Poorest communities are least resilient to weather extremes. *E&E Publishing, LLC*. Retrieved from http://www.businessinsurance.com/article/20140720/NEWS04/307209976?tags=%7C306%7C64%7C302

- Extreme Weather Hits Poorest Hardest

  - Basic Point: Similar to his other article on the vulnerabilities of the poor
  - Hulac, B. (2014, July 28). Extreme weather hits poorest hardest. *Scientific American*. Retrieved from http://www.scientificamerican.com/article/ extreme-weather-hits-poorest-hardest/?utm_medium=twitter&utm_ source=twitterfeed

Hurricane Hazards

- Florida Lawyer Says Hurricane Shutters Should Have Escape Latch

  - Basic Point: Buildings need to have preventative items, like shutters, with latches and locks that are accessible to the residents within the house to prevent deaths and insurance issues
  - Kleinberg, E. (2014, July 24). Florida lawyer says hurricane shutters should have escape latch. *Claimsjournal.com*. Retrieved from http://www.claims-journal.com/news/southeast/2014/07/24/252256.htm

- Miami-Dade must be proactive in fighting sea-level rise

  - Basic Point: The Miami-Dade area is prone to disasters due to sea-level rise and must become proactive in their attempts to combat before falling more below sea level

Bague, I. (2014, June 28). Miami-Dade must be proactive in fighting sea-level rise. *Miami Herald*. Retrieved from http://www.miamiherald.com/2014/06/28/4206630/ miami-dade-must-be-proactive-in.html

Index

© Springer International Publishing Switzerland 2015

F.I. Rivera, N. Kapucu, *Disaster Vulnerability, Hazards and Resilience*, Environmental Hazards, DOI 10.1007/978-3-319-16453-3

Printed by Printforce, the Netherlands